Blackstone's

Police Investigator's
Distance Learning Workbook

Blackstone's

Police Investigator's Distance Learning Workbook

2005

John Ridout

OXFORD
UNIVERSITY PRESS

OXFORD
UNIVERSITY PRESS

Great Clarendon Street, Oxford OX2 6DP

Oxford University Press is a department of the University of Oxford.
It furthers the University's objective of excellence in research, scholarship,
and education by publishing worldwide in

Oxford New York

Auckland Bangkok Buenos Aires Cape Town Chennai
Dar es Salaam Delhi Hong Kong Istanbul Karachi Kolkata
Kuala Lumpur Madrid Melbourne Mexico City Mumbai Nairobi
São Paulo Shanghai Taipei Tokyo Toronto

Oxford is a registered trade mark of Oxford University Press
in the UK and certain other countries

Published in the United States
by Oxford University Press Inc., New York

A Blackstone Press Book

ISBN 0-19-927539-4

10 9 8 7 6 5 4 3 2 1

Typeset by Newgen Imaging Systems (P) Ltd., Chennai, India
Printed in Great Britain
on acid-free paper by
Ashford Colour Press Limited, Gosport, Hampshire

Contents

How to use *Blackstone's Investigator's Distance Learning Workbook 2005*

The Workbook is designed to assist you in studying the syllabus defined for Phase 1 of the ICIDP and acts as an introduction to the text contained within the *Investigator's Manual*.

The Workbook aims to further your knowledge and understanding of the topics covered and uses a variety of exercises to test your learning.

The questions and exercises that appear in this Workbook do not necessarily reflect the style of the questions that appear in the NIE.

The National Investigators Examination (NIE) contains questions based solely on the text within the *Investigator's Manual*.

Any feedback regarding content or editorial matters in this Workbook can be emailed to police.uk@oup.com.

PART ONE

Property Offences

Theft

Objectives

With regard to offences of theft and robbery (ss. 1 to 8, Theft Act 1968), at the end of this section, from a given set of circumstances, you will be able to:

1. Identify the points to prove for the offence of theft.
2. Distinguish between the '*mens rea*' and '*actus reus*' of theft.
3. Recognise when an appropriation would not be dishonest.
4. Recognise when an appropriation may be dishonest.
5. Apply the test for dishonesty in the case of *R* v *Ghosh*.
6. Identify what is an 'appropriation'.
7. Distinguish between the different types of 'property'.
8. Identify when property 'belongs to another'.
9. Identify an 'intention to permanently deprive'.
10. Identify the points to prove for the offence of robbery.

Introduction

A large amount of your time as an investigator will be taken up with allegations of theft. Many other offences, apart from pure theft, involve an element of stealing. In order to understand all these offences you must familiarise yourself with the **Theft Act 1968**. Start now by trying the activity that follows:

Through various stages of your training you would have covered the definition of theft. In the space provided, write down this definition.

....

....

....

....

....

....

The answer should not have caused you any difficulty. Have a look at the definition below and see how close you were.

Section 1, Theft Act 1968

A person is guilty of theft if he dishonestly appropriates property belonging to another with the intention of permanently depriving the other of it; and 'thief' and 'steal' shall be construed accordingly.

If you were unable to write this out, in full, do not go on until you can. This definition is so important to the work of an investigator of crime, that you must know it.

 There are five points to establish in an offence of theft, and they are all in the definition. Please list them below:

....

....

....

....

....

The five points that make up the definition of theft are:

A person is guilty of theft if he or she:

(a) dishonestly

(b) appropriates

(c) property

(d) belonging to another

(e) with the intention of permanently depriving the other of it.

Each of the points from (a) to (e) are important in their own right and will be dealt with later in this section.

You may be wondering what the last part of the definition means, i.e. 'and "thief" and "steal" shall be construed accordingly'. As we go through the Theft Act 1968 these words will appear. 'Thief' describes the person who commits theft and 'steal' the act of theft.

Generally, to commit a crime, a person must not only do an act contrary to law, but also have a guilty mind.

| Guilty thought | + | Guilty deed | = | Crime |

In law the guilty thought or mind is known as the 'mens rea' and the guilty act or deed is known as the 'actus reus'.

| Mens rea | + | Actus reus | = | Crime |

Complete the table by identifying which of the points are linked to '*mens rea*' and which are linked to '*actus reus*'. Indicate in the appropriate column with a tick.

Points to prove theft	Mens rea	Actus reus
(a) dishonestly		
(b) appropriates		
(c) property		
(d) belonging to another		
(e) with the intention of permanently depriving the other of it		

Mens rea

The guilty thought. The state of mind needed by the offence, e.g. intention to commit an act, recklessness, dishonesty.

Actus reus

Any unlawful act (or, on occasions, an omission).

Before we move on to the next section there is a second subsection to s. 1.

Section 1(2), Theft Act 1968

It is immaterial whether the appropriation is made with a view to gain, or is made for the thief's own benefit.

In other words the thief does not need to gain or benefit from the theft. The thief can in fact throw the property away.

Let us now look in more detail at the component parts of theft, i.e. the points to prove.

Your table (above) should look like this:

Points to prove theft	Mens rea	Actus reus
(a) dishonestly	✓	
(b) appropriates		✓
(c) property		✓
(d) belonging to another		✓
(e) with the intention of permanently depriving the other of it	✓	

You should notice that the points indicated in the '*mens rea*' column are both states of mind, i.e. a dishonest mind and an intention in the mind. Anything that does not fall within a person's state of mind is related to the '*actus reus*' of the offence.

Dishonesty

What do you think 'dishonesty' means?

Spend a few minutes considering the list below, then tick which, if any, set of circumstances in your opinion is/are dishonest.

Circumstances	Could be dishonest	Would not be dishonest
(a) A woman finds £1 coin in the street and keeps it in the belief that the owner could not be found by taking reasonable steps.		
(b) An old man keeps a football that has been kicked into his garden by some children. He believes that he has a right in law to keep it.		
(c) A man takes his neighbour's pint of milk leaving the price of it, even though he realises his neighbour needs the milk.		
(d) A youth goes up behind a woman in a bus queue and seeing her purse in her open shoulder bag takes the purse and runs off with it.		
(e) Having cut his hand at work, a shop assistant takes some cash out of the till to buy a bandage, believing that the shop owners would have consented had they known of the circumstances.		
(f) An old lady selects a bacon joint from a display in a supermarket and then leaves the shop without paying.		

The answer to this activity can be found in s. 2, Theft Act 1968, which deals with the meaning of dishonesty. Look at the definition below.

Section 2, Theft Act 1968

(1) A person's appropriation of property belonging to another is not to be regarded as dishonest—

    (a) if he appropriates the property in the belief that he has in law the right to deprive the other of it, on behalf of himself or of a third person; or

    (b) if he appropriates the property in the belief that he would have the other's consent if the other knew of the appropriation and the circumstances of it; or

    (c) (except where the property came to him as trustee or personal representative) if he appropriates the property in the belief that the person to whom the property belongs cannot be discovered by taking reasonable steps.

(2) A person's appropriation of property belonging to another may be dishonest notwithstanding that he is willing to pay for the property.

Your table should have looked as below. The last column has been added to show you what part of s. 2 the answers came from.

You can see below that circumstances (a), (b) and (e) fall within the parts of s. 2 that tell you when something *is not* dishonest, whereas (c) comes within the part that tells you when something *may be* dishonest. Section 2(1) of the Theft Act 1968 is peculiar in that instead of giving a firm definition of what is dishonest, it only tells us really what will not be considered dishonest. Section 2(2) tells you when an act may be dishonest in a particular instance. What about (d) and (f)? You might think that in those circumstances (d) is clearly dishonest, whereas in (f) you may have thought you were given too little information, making a decision on dishonesty impossible. In any event, when all the facts are known, a jury or magistrate must decide on dishonesty.

Where s. 2 is not applicable in a particular prosecution, the jury or magistrate will need to consider the every day meaning of dishonesty. To help them, the jury or magistrate will be asked to apply the guidance in *R* v *Ghosh* [1982] QB 1053. They will need to ask the following questions in the following sequence:

(a) Were the actions of the defendant dishonest, according to the ordinary standards of reasonable and honest people?

If the jury or magistrate do consider the actions of the defendant to be dishonest, they should ask the second question.

(b) Did the defendant realise that what he or she did was 'dishonest' *by those standards*?

The jury or magistrate will always decide whether an act is dishonest or not, but if they think that the defendant really believed that what he or she did was acceptable, then they should acquit that defendant. Therefore, this amounts to a subjective test for dishonesty.

If you want to read more about dishonesty, see the section on dishonesty in the **Investigator's Manual**.

Circumstances	Could be dishonest	Would not be dishonest	Relevant section or decision
(a) A woman finds £1 coin in the street and keeps it in the belief that the owner could not be found by taking reasonable steps.		✓	s. 2(1)(c)
(b) An old man keeps a football that has been kicked into his garden by some children. He believes that he has a right in law to keep it.		✓	s. 2(1)(a)
(c) A man takes his neighbour's pint of milk leaving the price of it, even though he realises his neighbour needs the milk.	✓		s. 2(2)
(d) A youth goes up behind a woman in a bus queue and seeing her purse in her open shoulder bag takes the purse and runs off with it.	✓		*R* v *Ghosh*
(e) Having cut his hand at work, a shop assistant takes some cash out of the till to buy a bandage, believing that the shop owners would have consented had they known of the circumstances.		✓	s. 2(1)(b)
(f) An old lady selects a bacon joint from a display in a supermarket and then leaves the shop without paying.	✓		*R* v *Ghosh*

Appropriates

Have a look below at the definition of appropriates as given by s. 3(1), Theft Act 1968. Write down in the space provided, what you consider to be the key phrase that sums up the meaning of appropriation.

....

....

....

....

Section 3(1), Theft Act 1968

Any assumption by a person of the rights of an owner amounts to an appropriation, and this includes, where he has come by the property (innocently or not) without stealing it, any later assumption of a right to it by keeping or dealing with it as owner.

You should have spotted that the very first phrase—'any assumption by a person of the rights of an owner amounts to an appropriation'—is the basis of appropriation. The rest of the definition only extends the meaning.

It goes on to say that theft can include the situation where a person dishonestly retains or disposes of property after an acquisition that was in fact originally innocent.

There is a second part of the definition of appropriation which comes in s. 3(2):

Section 3(2), Theft Act 1968

Where property or a right or interest in property is or purports to be transferred for value to a person acting in good faith, no later assumption by him of rights which he believed himself to be acquiring shall, by reason of any defect in the transferor's title, amount to theft of the property.

This section provides protection for the innocent purchaser of stolen property who, but for s. 3(2) would be guilty of theft should they decide to hold on to it. However, the innocent purchaser will never become the owner of the property. Title to stolen property usually remains with the person who owned the property before the theft.

Ever since the Act was introduced in 1968, the definition of 'appropriates' has caused legal argument.

The case of *R* v *Gomez* [1993] AC 442 has helped to make this part of the law clearer. In this case the House of Lords decided that there are occasions where property can be appropriated even though the owner has consented or given authority to the thief to take possession or control of the property. You can read the full commentary in your **Investigator's Manual**.

Property

Section 4, Theft Act 1968 defines property. The section is divided into four sub-sections, but we will only be dealing with the first of these here. Have a look at it below.

Section 4(1), Theft Act 1968

'Property' includes money and all other property, real or personal, including things in action and other intangible property.

Look at the petrol station below and identify at least one example of each type of property from s. 4(1). Make a note of your choices in the table below.

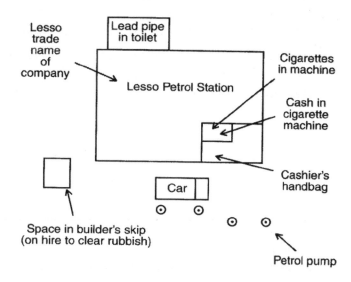

Put your answers in the table below.

Money	Real	Personal	Things in action	Other intangible property

The answers are listed in the table at the bottom of the page.

'Things in action' includes patents and trademarks and other property that can be enforced only by legal action. They do not exist in a physical form. This could include the contents of a bank account where the account is in credit or within the limits of an overdraft facility. 'Other intangible property' would include computer software programmes and perhaps credits accumulated on a smart card.

Section 4(2) tells us that you cannot steal land, but then goes on to give four exceptions to this rule. You will find these in your **Investigator's Manual**.

Money	Real	Personal	Things in action	Other intangible property
Cash in cigarette machine	Petrol pump	Cigarettes in machine	Trade name of company	Space in builder's skip
	Petrol station sign	Cashier's handbag		
	Lead pipe in toilet	Car		

Belonging to Another

Look at the following set of circumstances which concern a motor car. Identify which people have possession or control or have any proprietary right or interest in the car. In other words, who can it be stolen from?

Amnick Gill has bought a BMW motor car on a three-year hire purchase agreement with money from the Odeon Trust Finance Company. A mechanical fault develops in the first 12 months and his son drives the car to a local garage. Having completed the repair, the garage owner has the car road tested by one of the mechanics.

....

....

....

....

The answer is that the car could have been stolen from any of the people mentioned in the scenario: Amnick Gill and the Finance Company both have a proprietary right of interest in the car; the son, the garage owner and the mechanic all have possession or control of it at some stage.

Incidentally, had the car been bought with a bank loan, the bank would not have a proprietary interest in the car. A bank loan provides the purchaser with the money to buy the car and any default in payment gives the bank no automatic right to recover the car. Hire purchase companies have certain rights to repossess the car when repayments are not kept up-to-date.

It is possible to steal your own property. For instance, consider the situation where a garage makes repairs to a car and a dispute over the costs arises with the owner. Where the owner of the car acts dishonestly by returning to the garage out of business hours and takes the car back, the owner could be stealing their own car from the possession and control of the garage.

The definition in s. 5(1) below explains this.

Section 5(1), Theft Act 1968

Property shall be regarded as belonging to any person having possession or control of it, or having in it any proprietary right or interest (not being an equitable interest arising only from an agreement to transfer or grant an interest).

You will know that s. 5 actually includes four other sub-sections. These refer to trusts, property received under a certain obligation, property appropriated by mistake and property belonging to a corporation.

To find out more, you can read about 'belonging to another' in your **Investigator's Manual**.

Intention to Permanently Deprive

Section 6, Theft Act 1968

(1) A person appropriating property belonging to another without meaning the other permanently to lose the thing itself is nevertheless to be regarded as having the intention of permanently depriving the other of it if his intention is to treat the thing as his own to dispose of regardless of the other's rights; and a borrowing or lending of it may amount to so treating it if, but only if, the borrowing or lending is for a period and in circumstances making it equivalent to an outright taking or disposal.

(2) Without prejudice to the generality of subsection (1) above, where a person, having possession or control (lawfully or not) of property belonging to another, parts with the property under a condition as to its return which he may not be able to perform, this (if done for purpose of his own and without the other's authority) amounts to treating the property as his own to dispose of regardless of the other's rights.

Having had a look at the full definition, you will see that there are four different ways of 'intending to permanently deprive' included within it. Identify these four ways by picking out the key words from the definition and describing each of them below.

....

....

....

....

You should have identified the following as the key words for each of the four ways:

(a) An intention to permanently deprive.

(b) An intention to treat the thing as your own to dispose of regardless of the other's rights.

(c) A borrowing or lending equivalent to an outright taking or disposal.

(d) Parting with other's property under a condition as to its return which you may not be able to perform.

An 'intention to permanently deprive' speaks for itself and, in the definition, gets no explanation at all. Case law has gone to considerable lengths to explain that the purpose of s. 6 is to clarify, but not restrict, the meaning of 'intention to permanently deprive'. Therefore, the definition is written in such a way as to give illustrations of what amounts to a dishonest intention, but without being restrictive. The definition actually tells us about those times when an intention to permanently deprive may not appear so straightforward.

Below you are given four examples of 'intention to permanently deprive' covering each of the four ways listed above. Identify which example falls within each category by matching the letters to the numbers below.

Ways of tending to permanently deprive	Examples
1. An intention to permanently deprive.	(a) Biggs enters a scrap merchant's yard and takes a load of lead. He returns later to the yard and sells the lead to the unsuspecting dealer.
2. An intention to treat the thing as his own to dispose of regardless of the other's rights.	(b) Fraser takes cash from a wallet in the changing rooms and puts it in his bank account.
3. A borrowing or lending equivalent to an outright taking or disposal.	(c) Jones, a cashier, borrows cash from the till and places a bet on a horse, intending to replace the money later.
4. Parting with another's property under a condition as to its return which he may not be able to perform.	(d) Cooper is given a season ticket to a football club for one match but kept it until the end of the season (20 matches). He then gives the season ticket back.

(1) . . . . . . (2) . . . . . . (3) . . . . . . (4) . . . . . .

You should have decided that Biggs fell within category 2, Fraser in 1, Jones in 4 and Cooper in 3. These were just one example of each.

Let us now deal with an actual allegation of theft. Identify the five elements of theft from the circumstances given below.

Case Study 1

Here is a reported allegation of theft, which we will be asking you to refer to from time to time. The case will develop as the investigating officer makes relevant enquiries.

Investigator Crouch was on duty when the manager of a local jewellers reported the theft of a diamond ring worth £15,000. The particulars of the theft were recorded on a crime report. The crime was allocated to Investigator Crouch and here is the crime report.

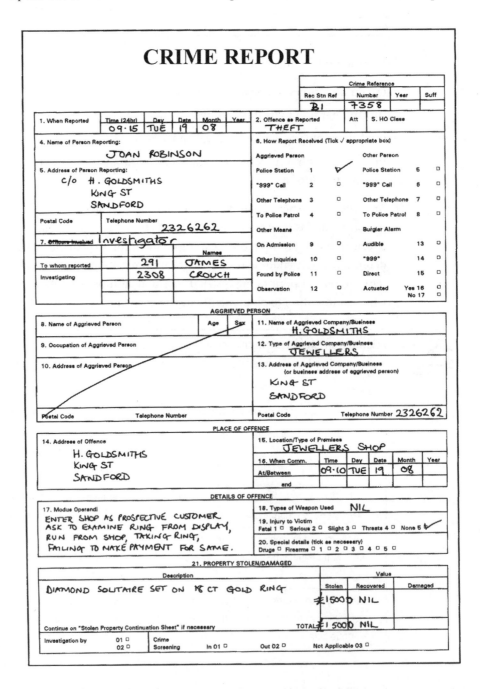

Look at the crime report above and identify how many of the elements of theft appear to be present. Do this by explaining how the circumstances satisfy the elements of theft.

Element of theft	How the circumstances of the case study satisfy the element of theft
Dishonesty	
Appropriation	
Property	
Belonging to another	
Intention to permanently deprive	

Your answer should have looked like this.

Element of theft	How the circumstances of the case study satisfy the element of theft
Dishonesty	By running off with the ring the suspect shows signs of dishonesty.
Appropriation	The suspect has assumed some rights of the owner by taking it out of the shop.
Property	The ring is personal property.
Belonging to another	The ring clearly belongs to H Goldsmith, Jewellers, and is in the possession of the manager Joan Robinson.
Intention to permanently deprive	By running off and giving no explanation, leaving no address or making no arrangement for the return of the ring, the suspect indicated, but not conclusively so, an intention to permanently deprive.

The 'actus reus' elements of the offence ('appropriates', 'property' and 'belonging to another') are often straightforward, because these are concrete parts of theft that will either exist or have occurred.

The 'mens rea' is not always so obvious because it is a state of mind. Although you may have identified 'dishonesty' and 'intention to permanently deprive' from the circumstances, you will never be completely sure until you have fully investigated the case including interviewing the suspect and addressing any defences or mitigating reasons, because these elements are in the mind of the suspect.

Investigator Couch attended the scene at 9.45 that morning and spoke to the manager Joan Robinson. She directed Investigator Crouch to speak to one of her assistants, Peter McKay, who had been serving the suspect.

Have a look at the version of events given to Investigator Crouch by McKay. In the space provided state what offence(s) may have been committed.

'He came in and asked to see a ring which was on display in the window. I brought the ring pad and took the ring from it that he wanted to see. I handed him the ring and he looked at it then held it up as if to catch the light and turned as he did so and began to run from the shop. Fortunately I wasn't behind the counter and immediately ran after him and caught him by the arm in the shop doorway. I began to pull him back into the shop but he was struggling. The next thing I know he punched me in the face knocking me to the floor. He then ran off with the ring. But in the struggle he dropped a cheque book on the floor.'

Offences: ....

....

....

....

....

....

....

Your answer would probably have included theft and an assault, and may also have mentioned robbery. Let us now have a look at how robbery fits in with theft.

Robbery

Have a look at the definition of robbery in s. 8, Theft Act 1968 and try the activity below.

Section 8(1), Theft Act 1968

A person is guilty of robbery if he steals, and immediately before or at the time of doing so, and in order to do so, he uses force on any person or puts or seeks to put any person in fear of being then and there subjected to force.

From the definition, fill in the empty boxes which will then show the definition broken down into its five elements.

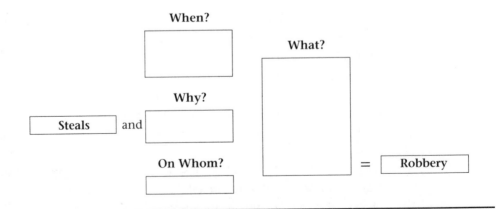

Your diagram should have looked like the one below.

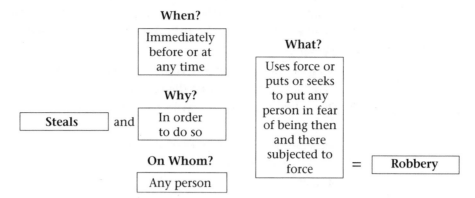

Note that 'steals' means theft only as defined by s. 1, Theft Act 1968.

Robbery is an aggravated form of theft and proof of the commission of the theft is essential to secure a conviction for robbery. However, the property being stolen does not need to be in the immediate possession of the person to whom the threat is made or upon whom the force was used. For example, a bank manager may be taken to a bank by a thief to collect money while an accomplice holds the family of the manager hostage under the threat of assault.

Let us now go back to Case Study 1. You will remember that the suspect had taken the ring and used force on the shop assistant only to aid escape. In these circumstances the offences of theft and assault are clearly present. There appears to be no robbery because the force was not used immediately before or at the time of the stealing, nor in order to steal the ring. This will in fact depend upon the interpretation of 'appropriates'.

Appropriation can be regarded as a continuing act. Therefore, where the suspect is still in the process of appropriating the property when an assault takes place, robbery may be the most appropriate charge. Practically, it may be advisable to charge both theft and robbery as alternatives.

Summary

As you have seen, theft and robbery may appear simple subjects to understand but can be complex areas of the law. To help you understand the subject better you should read the section on theft in your **Investigator's Manual**.

Investigator Crouch has so far made little progress in identifying the suspect in the theft. Investigator Crouch has decided that the best line of enquiry will be to investigate the circumstances surrounding the cheque book dropped in the jeweller's shop by the suspect.

Investigator Crouch in fact finds out that the cheque book was stolen in a burglary at 23 Elmwood Road, Sandford, a few days earlier.

Self-assessment Test

Having completed this section, test yourself against the objectives outlined at the beginning of the section. You will find the answers below.

1. At lunchtime Davies was walking through a city centre car park and saw a laptop computer on the back seat of a car. Davies realised that the computer was valuable, as they usually cost several hundred pounds. Davies continued to walk past the car, looking

around for anybody who may have been watching. Once satisfied that it was safe to approach the car, Davies tried the driver's door and found it unlocked. Davies took one more look around and reached into the car and quickly took the computer and walked away towards the exit and left the car park. Davies immediately sold the computer to a handler for £50.

With regard to s. 1, Theft Act 1968, which of the following statements is correct?

(a) All of the points to prove for theft appear to be present in the circumstances provided.

(b) The furtive behaviour of Davies would form part of the '*actus reus*' of the offence of theft.

(c) The laptop computer is an example of property that is a 'thing in action'.

(d) The appropriation of the laptop computer forms part of the '*mens rea*' of the offence of theft.

2. With regard to the '*mens rea*' for an offence of theft (section 1 and 2, Theft Act 1968), which of the following statements is correct?

(a) Black was on trial for theft of £10 from a cash till. Black gave evidence that he had taken the £10 from the cash till and believed he had a moral right to do so. Under these circumstances Black could be acquitted as a consequence of s. 2(1)(a).

(b) Frost was on trial for theft of a sports kit from a friend's kit bag. Frost gave evidence that, when she had taken the sports kit, she had believed that the friend would have consented had she known of the circumstances. Under these circumstances Frost could be acquitted as a consequence of s. 2(1)(b).

(c) Carrick was on trial for the theft of a car from a person who had the car up for sale. Carrick gave evidence that he had taken the car without telling the owner but had posted the right sale price through the owner's letterbox. Under these circumstances Carrick could be acquitted because willingness to pay will never be dishonest as a consequence of s. 2(2).

(d) Baldwin was a trustee to a charity and on trial for theft of funds from the charity. Baldwin gave evidence that, taking all reasonable steps, she had been unable to find the owner of the money. Under these circumstances Baldwin could be acquitted as a consequence of s. 2(1)(c).

3. Evans went into a computer software store, selected an expensive computer game and left without paying. With regard to assessing dishonesty in an offence of theft a jury must apply *R* v *Ghosh*. Consequently which of the following statements is correct?

(a) At the trial of Evans, the jury accepted that what Evans did was dishonest according to the ordinary standards of reasonable and honest people. The jury can convict on this criterion alone.

(b) At the trial of Evans, the jury accepted that what Evans did was dishonest according to the ordinary standards of reasonable and honest people. This amounts to a subjective test for dishonesty.

(c) At the trial of Evans, the jury accepted that what Evans did was dishonest according to the ordinary standards of reasonable and honest people and that Evans realised that walking out with the computer game was dishonest by those standards. The jury can convict on this criterion of dishonesty.

(d) At the trial of Evans, the jury accepted that what Evans did was dishonest according to *R* v *Ghosh*. This amounts to an objective test for dishonesty.

4. Roberts did not own a car and decided to get himself one and keep it. He went out looking for an Audi saloon car. Roberts quickly found one parked in a station car park, broke into it through the driver's door and drove off. Roberts took it to a lock-up garage and

immediately changed the registration plates to a number from a scrapped car of a similar description. After a short time Roberts forged some registration documents for the car and decided to sell it at a car auction. Patel bought it in good faith and paid a fair price for it. Patel retained the car after finding out that it was in fact stolen.

With regard to 'appropriation', (s. 3, Theft Act 1968), which of the following statements is correct?

(a) Roberts appropriated the car because he had at some stage assumed the rights of the owner.

(b) As Patel retained the car after finding out that it was in fact stolen, he would be guilty of theft in these circumstances.

(c) Roberts at no stage appropriated the car because he had committed the offence of taking a motor vehicle without the consent of the owner.

(d) Patel at no stage appropriated the car because he had committed the offence of taking a motor vehicle without the consent of the owner.

5. A film actor lives on an exclusive 20 acres estate surrounded by high walls with sophisticated electronic surveillance. The actor's initials 'HG' are well known and form a registered trademark for worldwide merchandising. The estate has a large garage containing a range of high value cars.

With regard to 'property' (s. 4, Theft Act 1968), which of the following statements correct?

(a) The trademark 'HG' is not 'property' for the purpose of theft and cannot be stolen.

(b) The cars are 'things in action' property and can be stolen.

(c) The space in the garage is 'real' and can be stolen.

(d) The electronic surveillance equipment is real property and can be stolen.

6. Stone purchased a personal stereo player from Dexon's Electrical using a bank loan with payments spread over 24 months. The player developed a fault after only three months and she took it back to the retailer under warranty. The retailer sent the player to the workshop of a local repairer. While at the workshop, the player was stolen.

With regard to 'belonging to another' (s. 5, Theft Act 1968), which of the following statements is correct?

(a) The player cannot have been stolen from Stone because it had passed out of her control.

(b) The bank has a proprietary right in the player and therefore can be a victim of the theft.

(c) The player is in the possession of the local repairer and can be stolen from them.

(d) The player can be stolen only from the company that supplied the warranty.

7. Hill knew that her friend, who worked in the city centre, had just bought a weekly travel card. The card had four days' travel still to run. Hill took the travel card from her friend's wallet and used it to travel until its expiry date. She then put it back in her friend's wallet.

With regard to 'intention to permanently deprive' (s. 6, Theft Act 1968), which of the following statements is correct?

(a) Hill 'intended to permanently deprive' because she parted with her friend's travel card under a condition as to its return which she may not have been able to perform.

(b) Hill borrowed the travel card and this was equivalent to an outright taking.

(c) Hill could not have intended to permanently deprive her friend of the travel card because her friend got the property back.

(d) Hill always intended to return the travel card and cannot be guilty of theft under these circumstances.

8. Cole telephoned the manager of a local wine bar. Cole said: 'Put £1,000 in a carrier bag and leave it in the telephone box on the corner of York Street. If you don't do this

immediately I will be around to sort you out tomorrow. What ever you do don't call the police'. The manager was not concerned about this threat.

With regard to 'robbery' (s. 8, Theft Act 1968), which of the following statements is correct?

(a) All the points to prove for robbery are present in the circumstances.

(b) For robbery to be committed, such a threat to subject someone to force can never be delivered by telephone.

(c) Robbery was not committed because force is not used and nobody was at that time put in fear of being then and there subjected to force.

(d) Robbery was not committed because the manager was not concerned about the threat.

Answers to Self-assessment Test

1. Answer (a) is correct because although behaviour is generally part of the '*actus reus*', the furtiveness' of the behaviour would tend to prove dishonesty and this forms part of the '*mens rea*'. The computer is personal property and appropriation forms part of the '*actus reus*'.

2. Answer (b) is correct because s. 2(1)(a) refers to a legal not moral right. Willingness to pay can be dishonest under s. 2(2). A trustee is an exception to s. 2(1)(c).

3. Answer (c) is correct because both criteria need to be met in the correct order. The first criterion outlined in the case amounts to an objective test but becomes a subjective test by virtue of the second criterion.

4. Answer (a) is correct because Patel is protected by s. 3(2). Taking the car without consent does not affect s. 3 'appropriates' in theft.

5. Answer (d) is correct because the trademark is a 'thing in action' and can be stolen. The cars are personal property and the space 'other intangible property'.

6. Answer (c) is correct because the property can still be stolen from the owner even though it is not in their control. A bank, in these circumstances, does not have a proprietary right in the property in the same way that a hire purchase company would.

7. Answer (b) is correct because there was no parting with the property under a condition as to its return.

8. Answer (c) is correct because there was nobody *there* and *then* put in fear. Such a threat could be made by telephone as long as some person is in fact having force applied to them or the threat of its immediate application.

Blackmail

Objectives

With regard to offence of blackmail (ss. 21 and 34, Theft Act 1968), at the end of this section, from a given set of circumstances, you will be able to:

- Identify the points to prove for the offence of blackmail.
- Distinguish between the offences of robbery and blackmail.
- Recognise a 'gain' or 'loss' in the offence of blackmail.

Introduction

To the investigating officer, the offence of blackmail is an area of the law which can be both confusing and complex. Without a knowledge of the subject, difficulty may be experienced in identifying the offence from any given circumstances and in some cases, distinguishing it from the offence of robbery.

Case Study 2

Thieves broke into the house of a high-profile politician and best-selling author. Whilst in the house, they found the author's briefcase which, amongst other things, contained a number of photographs of him taking part in sexual activities with children. Realising their potential value, they took the briefcase with them.

The author returned home the following day and discovered the burglary. Particularly concerned over the loss of the photographs, and understandably anxious to seek their return, he decided not to inform the police. Instead, he placed an advertisement in the local newspaper appealing to the thieves for the return of his briefcase and contents, with the promise that no questions would be asked.

Advertising for Goods Stolen or Lost

You will probably have realised that, in placing such an advertisement, the author committed an offence under s. 23, Theft Act 1968. Briefly, this section creates an offence for a person advertising, **and** any person who prints or publishes the advertisement, for the return of any goods which have been stolen or lost using any words to the effect that no questions will be asked or that the person producing the goods will be safe from apprehension or inquiry, or that any money paid for the purchase of the goods or advanced by way of loan on them will be repaid.

However, on reading the advertisement, the thieves recognised the chairman's vulnerability, and decided to profit from it and send him this letter:

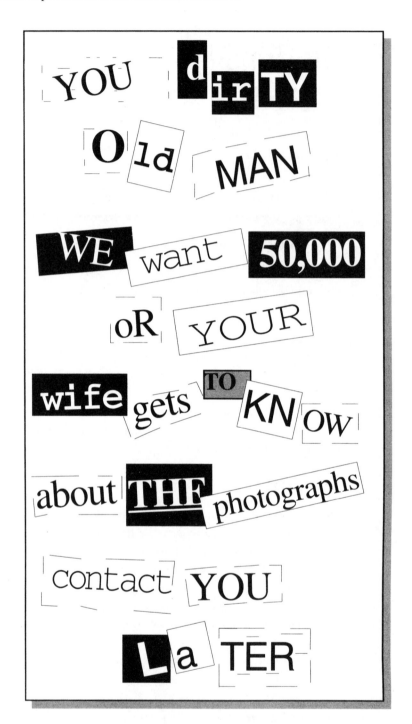

Blackmail

Are they guilty of blackmail?

Let's examine the letter and compare it stage by stage with this definition:

Section 21, Theft Act 1968

A person is guilty of blackmail if, with a view to gain for himself or another, or with intent to cause loss to another, he makes any unwarranted demand with menaces; and for this purpose a demand with menaces is unwarranted unless the person making it does so in the belief (a) that he has reasonable grounds for making the demand and (b) that the use of the menaces is a proper means of reinforcing that demand.

So, certain points must be established in order to prove an offence of blackmail. In every case we must ask ourselves a few basic questions. Those questions are set out below. Read them and for each provide the Answer 'yes' or 'no'.

Question	Answer
1. Was the letter sent with a view to gain or cause a loss?	
2. Does it contain a demand?	
3. If so was that demand unwarranted?	
4. Does a menace exist?	

The answer to the first question seems obvious. Anybody reading that letter would undoubtedly come to the conclusion that it was sent with a view to gaining the sum of £50,000.

But does it contain a demand? The words 'we want £50,000' is quite clearly a demand.

But what if the letter was couched in the following terms?

Dear Sir

We were delighted to obtain your very interesting photographs. We would be grateful if you would, in accordance with instructions you will receive later, forward to us the sum of £50,000.

If you cannot see your way clear to accede to this request, it will be our painful duty to furnish your wife with copies of the photographs.

Yours faithfully

'Demand' seems a harsh word to describe such a polite request. But case law tells us that even the most courteous request will be considered a 'demand' if it is accompanied by a menace.

So was the demand unwarranted? There would seem to be no justification for the thieves to seek payment of money in this way. We can only conclude, therefore, that their demand was unwarranted.

Finally, we must establish whether or not a menace existed. A menace has been described in this context as 'a declaration to inflict evil or injury' and includes 'threats of any action detrimental or unpleasant to the person addressed'. The words 'or your wife gets to know about the photographs' fits nicely into that description. One can safely view that as a 'menace'.

R v Clear [1968]

It has been decided that provided the menace is made, it is not necessary that the proposed victim is actually alarmed but only that a person of normal stability and courage might be influenced or made apprehensive.

If your answer to each of the four questions was 'yes', the letter, in its entirety, amounts to blackmail. In the unlikely event that one or more of your answers was 'no', then blackmail cannot be proved.

'Gain' and 'Loss'

You will have noticed the use of these words within the definition. In every case, it in necessary to prove that the blackmailer made his demand with a view to 'gain for himself or another' or intend to 'cause loss to another'.

'With a view' is a unique legal term different from 'intent'. 'With a view' appears to mean a lesser state of mind than intending something to happen. It is more like 'contemplating something that realistically might occur'. So the thieves in our example were making a demand expecting to make a gain of money. Had the thieves made a demand so that the author would make a loss the thieves would need the higher state of mind of 'intending' the author to make the loss (*R* v *Zaman* [2002]).

Obviously, to gain is usually the purpose of blackmail. But it is possible that, occasionally, a person might threaten something unpleasant if the intended victim does not, for example, destroy some documents or abandon a claim, thereby causing a loss to another.

To test your interpretation of 'gain' and 'loss', try the following activity.

In which, if either, of the following circumstances has the offence of blackmail been committed?

1. A window cleaner cleans the windows of a house by agreement with the occupant. The occupant than refuses to pay for the work saying that, if he insists on payment, she will allege to the neighbours that he indecently assaulted her.

2. A shop assistant becomes aware that his female colleague has taken time off work by falsely stating she was ill. He tells her that he will inform the Manager unless she agrees to have sexual intercourse with him.

(a) 1 only   (c) both

(b) 2 only   (d) neither

Answer: ....

In both instances a demand has been made, and in each there exists a menace. What's more, in each case, it could be said, someone stands to gain and, indeed, someone stands to lose. But do they both amount to blackmail?

Let's look now at what the legislators mean by their use of the words 'gain' and 'loss'.

Section 34(2)(a), Theft Act 1968

For the purposes of this Act:

(a) 'Gain' and 'loss' are to be construed as extending only to gain or loss in money or other property, but as extending to any such gain or loss whether temporary or permanent; and—
  (i) 'gain' includes a gain by keeping what one has, as well as a gain by getting what one has not; and
  (ii) 'loss' includes a loss by not getting what one might get, as well as a loss by parting with what one has.

Now we can see that, since 'gain' and 'loss' refer only to a gain or loss of money or other property, the answer to our question is (a). In the second scenario the contemplated gain was that of sexual intercourse, which cannot be construed as property, and, whilst other offences may exist, blackmail does not.

It is also worth remembering that the 'gain' or 'loss' need not necessarily be permanent. So, such comment as 'If you don't let me use your car for the weekend, I'll tell everyone at the club you're homosexual', could well amount to blackmail, even though the contemplated 'loss' or 'gain' is temporary.

You may have thought the wording at (i) and (ii) in the definition a little awkward. All it's really saying is that there are two ways of making a 'gain' and two ways of incurring a 'loss'.

We've discussed one or two examples of when a person may be guilty of blackmail. Now let's look at the times when he would not be guilty.

Defence

Earlier, when we were looking at the definition of blackmail, you probably realised that no mention was made of the exceptions contained in it. Look at the second part of the definition:

Section 21(1), Theft Act 1968

... and for this purpose a demand with menaces is unwarranted UNLESS the person making it does so in the belief:

(a) that he has reasonable grounds for making the demand,

AND

(b) that the use of the menaces is a proper means of reinforcing that demand.

Consider the following statements:

'If you don't pay the money you owe me, I'll arrange for you to be served with a writ.

'If you don't keep up your payments on the car, I'll come round and re-possess it.'

Both the statements contain a demand, both are coupled with a menace and what's more both are frequently made in everyday life. However, because those making them do so in the honest belief that they have:

(a) reasonable grounds for making the demand; and

(b) the use of the menace is a proper means of re-enforcing that demand,

the demand is not unwarranted and quite clearly, they commit no offence.

If, on the other hand, the statements were:

'If you don't pay the money you owe me, I'll arrange for you to be beaten up.'

'If you don't keep up your payments on the car, I'll come round and throw a brick through your window.'

Whilst those making the statements still have an honest belief that they have reasonable grounds for making the demand, they cannot possibly claim that the menace used was a proper means of reinforcing that demand. The offence of blackmail would, therefore, be complete.

The whole question hinges on whether or not a person had honest belief in the propriety of his demand **and** the menace to reinforce it.

Although criminal attempts are dealt with elsewhere, it is worth remembering that the courts have decided there is no such offence as attempted blackmail (*R* v *Moran* [1952]).

The offence is complete once the demand is made. With a letter, this is when it is posted in the public mail system.

Summary

This has been a brief insight into blackmail. If you remember that the essence of the offence is in a person making a demand for something with menaces knowing, *either* that he has no right to make the demand, *or* that the use of the menace is improper, then you won't go far wrong.

By breaking up the definition into its four parts and asking the questions we did earlier, you should identify when the offence is present.

Self-assessment Test

Having completed this chapter, test yourself against the objectives outlined at the beginning.

1. When interpreting the meaning of 'gain' and 'loss' in relation to blackmail, which, if either, of the following statements are correct?
   (i) Their meaning extends only to the gain or loss of money or other property.
   (ii) Their meaning extends only to a permanent gain or permanent loss.
   (a) (i) only   (c) both
   (b) (ii) only   (d) neither

2. When interpreting the meaning of 'gain' and 'loss' in relation to blackmail, which, if either, of the following statements are correct?
   (i) 'Gain' includes a gain by keeping what one has, as well as a gain by getting what one has not.
   (ii) 'Loss' includes a loss by not getting what one might get as well as a loss by parting with what one has.
   (a) (i) only   (c) both
   (b) (ii) only   (d) neither

3. A defence is written into the definition of blackmail. In which, if either, of the following circumstances could that defence be justifiably claimed?
   (i) An employee has genuinely earned overtime money, but his employer refuses to pay. He knows it is wrong, but threatens the employer with a gun, saying, 'If you don't pay me what I'm owed, I'll shoot you'.
   (ii) An employee genuinely believes he is owed overtime money, but his employer refuses to pay. He threatens the employer by saying, 'If you don't pay me what I'm owed, I'll start legal proceedings against you'.
   (a) (i) only   (c) both
   (b) (ii) only   (d) neither

4. Luke prepared an anonymous letter using his computer and printer. He intended to send the anonymous letter to Leslie. The letter said 'Unless you put £500,000 in used notes into a brief case and leave it outside the railway station I am going to tell the police about your unlawful business dealings'.
   With regard to the offence of blackmail (s. 21), at which of the following stages is the offence complete?
   (a) When Luke types the letter.
   (b) When Luke posted the anonymous letter to Leslie's home address by putting it into the post box at a post office.
   (c) When the anonymous letter passed through the letter box at Leslie's address.
   (d) When Leslie opened the anonymous letter and read the contents.

Answers to Self-assessment Test

1. Answer (a). The gain or loss must be property but can be a temporary gain or loss.

2. Answer (c). The definitions of 'gain' and 'loss' are included in the options provided.

3. Answer (b). The first employee needs to believe that they have reasonable grounds for making the demand and that the menace is the proper means of reinforcing it.

In this case the employee realises the threat is not the correct way of claiming the money owed.

4. Answer (b). No matter what the outcome of the demand may be, the offence is complete once the demand has been made. Where the demand is made by a letter the offence is complete as soon as the letter is posted (*Treacy* v *DPP* [1971]).

Burglary

Objectives

With regard to offences of burglary and aggravated burglary, (ss. 9 and 10, Theft Act 1968), at the end of this section, from a given set of circumstances, you will be able to:

1. Identify the points to prove for the offences of burglary.
2. Distinguish between burglary contrary to s. 9(1)(a) and s. 9(1)(b).
3. Identify the points to prove for the offences of aggravated burglary.
4. Distinguish between the different types of named articles mentioned in s. 10.
5. Apply the meaning of 'has with him' within the context of s. 10.

Burglary

In s. 9, Theft Act 1968, there are two types of burglary: s. 9(1)(a) and s. 9(1)(b). They differ in certain key respects, and we will deal with those differences soon, but both have three points in common.

From your memory of burglary, list below the common points for offences under s. 9(1)(a) and s. 9(1)(b).

....

....

....

....

....

The common points are 'enters', 'building or part of building' and 'as a trespasser'. In other words, no matter what type of burglary is committed, these three points must be present and proved.

Let us now look at the offences and how they differ.

Section 9(1)(a), Theft Act 1968

A person is guilty of burglary if—

(a) he enters any building or part of a building as a trespasser and with intent to commit any such offence as is mentioned in subsection (2) below ...

Section 9(2) lists the offences of stealing anything in the building or part of the building, inflicting grievous bodily harm on any person therein or doing unlawful damage to the building or anything therein.

Section 9(1)(b), Theft Act 1968

A person is guilty of burglary if—

...

(b) having entered into any building or part of a building as a trespasser he steals or attempts to steal anything in the building or that part of it or inflicts or attempts to inflict on any person therein any grievous bodily harm.

As you can see, the main difference between s. 9(1)(a) and s. 9(1)(b) is that s. 9(1)(a) requires an intention to commit one of the three offences listed in s. 9(2) only, while s. 9(1)(b) is the actual committing or attempting to commit one of two listed offences.

Look at the diagram below, which shows a street of houses. Assume those that are indicated have been entered by a suspect who was a trespasser. Decide from the circumstances what offence(s), if any, have been committed. Then tick the appropriate column on the next page.

In all cases below suspect enters as a trespasser.

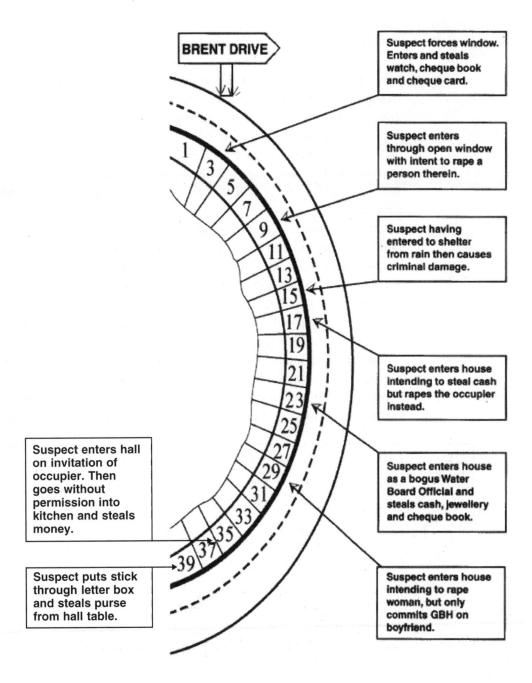

House number	Burglary s. 9(1)(a) committed	Burglary s. 9(1)(b) committed	No burglary
3			
9			
15			
17			
23			
29			
35			
39			

A s. 9(1)(a) burglary was committed at house 17. It is not clear from the information provided what intent was present at houses 3, 23, 35 and 39.

A s. 9(1)(b) burglary was committed at houses 3, 23, 29, 35 and 39.

No offence of burglary was committed at houses 9 and 15, although, the offence of criminal damage may have been committed at 15. Rape is not an ulterior offence. There is a specific offence of trespassing with intent to commit a sexual offence (s. 63 Sexual Offences Act 2003).

The buildings in the diagram are clearly buildings for the purpose of burglary. Buildings generally need to have a degree of permanence. Therefore, a boathouse will be a building. Similarly, a vehicle can be a building, when it is parked up and used as a dwelling and not as a vehicle for transportation.

Buildings are frequently divided up into different rooms or parts. Burglary can be committed when a person moves from one room or part to another as a trespasser. For example, you may enter the public hotel foyer as a bone fide guest, but as soon as you move into another part of the hotel that is private or restricted you would be satisfying the 'entering a building or part of a building' element of burglary.

Usually, 'entering' will be the movement of the whole body into a building. Case law shows that 'enter' means inserting any part of the body into a building to the slightest degree and includes inserting a simple or mechanical extension to the body. An example of the latter may be a stick or an extending grab device. For instance, where someone inserts their arm or a stick into a letterbox to drag out and steal some property, this would be 'enters' provided the 'entering' is to commit or attempt to commit one of the ulterior offences.

Below are the important differences in the two types of burglary:

Section 9(1)(a) burglary = *intent to*:

(a) steal; or

(b) inflict grievous bodily harm; or

(c) commit unlawful damage.

Section 9(1)(b) burglary = *actually*:

(a) steals or attempts to steal; or

(b) inflicts or attempts to inflict grievous bodily harm.

An important distinction between the two sections is that s. 9(1)(a) burglary is an offence of specific intent; s. 9(1)(b) burglary is an offence of basic intent. This is significant when intoxication is a factor in an investigation. Voluntary intoxication can be a defence to offences of specific intent, but not for offences of basic intent.

Aggravated Burglary

Introduction

The difference between burglary and aggravated burglary is simple. You know already what constitutes the offence of burglary. Aggravated burglary has exactly the same ingredients plus the additional evidence that, at the time of committing the offence, the accused had with them one or more of the four named articles listed below:

(a) firearm;

(b) imitation firearm;

(c) weapon of offence;

(d) explosive.

Therefore, a person is guilty of aggravated burglary.

Section 10(1), Theft Act 1968

A person is guilty of aggravated burglary if he commits any burglary and at the time has with him any firearm or imitation firearm, any weapon of offence, or any explosive ...

Or, in mnemonic form, has a WIFE:

Weapon of Offence

Imitation Firearm

Firearm

Explosive

Let us briefly discuss those named prohibited articles, because it is essential to understand their meaning.

Firearm

This has exactly the same meaning as provided by the Firearms Act 1968 but excludes references to component parts and accessories. For the purpose of aggravated burglary, therefore, a firearm is:

Section 57(1), Firearms Act 1968

... a lethal barrelled weapon of any description from which any shot, bullet or other missile can be discharged and includes—

(a) any prohibited weapon whether it is such a lethal weapon as aforesaid or not ...

You will see from the use of the words 'lethal barrelled' that, in addition to all the more obvious weapons, the term 'firearm' also includes within its meaning such weapons as shotguns, air guns and air pistols etc.

There is a common misconception that air weapons and shotguns are not firearms. This has probably come about because those types of guns can be lawfully possessed without the authority of a firearms certificate. But, in almost every case, such weapons are capable of causing death. Those that are so capable are, quite clearly, 'lethal barrelled' and, therefore 'firearms'.

Imitation Firearm

An imitation firearm is defined in the Theft Act 1968:

Section 10(1)(a), Theft Act 1968

... anything having the appearance of being a firearm, whether capable of being discharged or not ...

This definition obviously includes toy guns and replicas etc., but, under certain circumstances, it includes other articles as well.

The meaning of the words 'firearm' or 'imitation firearm' should cause you no difficulty. With your existing knowledge of the Firearms Act 1968, you will see there is very little difference. However, the remaining articles, may need more careful examination.

Weapon of Offence

Spend a few moments considering the list below and then select those articles which you think may properly be described as weapons of offence.

(a) sharpened metal comb;

(b) knuckleduster;

(c) length of rope;

(d) adhesive tape.

Answer: ....

Let us look at the definition as it appears in s. 10(1)(b).

Section 10(1)(b), Theft Act 1968

'weapon of offence' means any article made or adapted for use for causing injury to or incapacitating a person, or intended by the person having it with him for such use ...

You will have noticed that the definition is almost the same as that of an offensive weapon under the Prevention of Crime Act 1953. However, there is one important difference: the inclusion of the words 'or incapacitating a person' creates a significant extension.

Going back to our question. The knuckleduster is made to cause injury and is clearly a weapon of offence. The metal comb if sharpened for the specific purpose of causing injury, would then be adapted for that purpose and that too would be a weapon of offence. The remaining articles (rope and adhesive tape), although innocent in themselves, if carried with the intention to incapacitate (e.g. tying up and gagging), would be 'weapons of offence' for the purpose of aggravated burglary.

For your own benefit, spend a few minutes listing two or three further examples which would fit the following categories of 'weapons of offence'.

1. Made to cause injury...

2. Made to incapacitate...

3. Adapted to cause injury..

4. Adapted to incapacitate...

5. Intended to cause injury...

6. Intended to incapacitate..

You were probably able to bring any number of articles to mind for some categories. In others, you may have had difficulty in thinking of any. You will find that the ease or difficulty with which you were able to think of examples, usually corresponds to the frequency such articles will come to light during the course of your investigation.

Explosive

What sort of things come to your mind that could be described as explosives? List below as many items as you can, which you think fall within the meaning of 'explosive'.

....

....

....

....

....

Now check your list against the definition:

Section 10(1)(c), Theft Act 1968

'explosive' means any article manufactured for the purpose of producing a practical effect by explosion, or intended by the person having it with him for that purpose.

The word 'explosive', therefore, has a broad meaning. In addition to the more obvious items, such as semtex, it includes many other things, e.g. powerful public display firework.

Let us imagine for a moment that a burglar took with him a thunderflash. The purpose in doing so, would be to use it in creating a diversion in order to assist his escape should he be disturbed. Would he be guilty of aggravated burglary? The answer is surely yes.

Has with Him

You will have noticed the use of the phrase 'has with him' when we were distinguishing the offence. It is worthy of some explanation. Consider the following scenario:

Case Study 3

Three friends, Piper, Pickles and Collier, all experienced criminals who specialise in country house burglaries, decide to break into the home of the chief executive of a chain of grocery stores.

All three gain entry to a ground floor room in the house. Pickles has a handgun in a carrier bag. Pickles has not made, or attempted to make, any use of the gun. In fact, Pickles is rather hoping that he does not have to use it.

With that in mind, try the following activity.

Consider the following statements:

1. Pickles alone is guilty of aggravated burglary. Pickles is the only one to whom the phrase 'has with him' applies, since he alone has actual possession.

2. Provided the others know that Pickles has a gun, the phrase 'has with him' applies to them all. If such was the case, they are all guilty of aggravated burglary.

3. The gun was readily accessible to all three. The phrase 'has with him' therefore applies to them all whether or not the others knew that Pickles had a gun, and they are all guilty of aggravated burglary.

Which, if any, of the statements is correct?

(a) 1 only   (b) 2 only   (c) 3 only

Answer: ....

As indicated in the words used, the phrase 'has with him' is strictly limited in its interpretation, it means that there should be a requirement for a degree of immediate control, not necessarily actual physical possession.

Furthermore, there is a requirement to prove guilty knowledge, i.e. the accused person must knowingly 'have with him' the prohibited article.

The expression 'has with him' must be considered alongside 'at the time'. Where there is an entry with intent to commit one of the ulterior offences, the aggravated offence is committed on entry to the building or part of the building. This means that a person can enter a house unarmed and arm himself in the hall with a golf club and would not 'have the golf club with him' at that stage. But should he leave the hall and enter the lounge (another part of the building) he will 'have the golf club with him'.

However, where the person goes on to commit theft or grievous bodily harm, the 'has with him' will be at the time of the theft or grievous bodily harm. This means where a person is committing burglary in a kitchen and instantly picks up a knife and inflicts grievous bodily harm on the occupier, he will 'have the knife with him' for the purpose of aggravated burglary.

When, when two or more people commit burglary and one of them has physical possession of a prohibited article, then all who knew of the article are guilty of aggravated burglary.

Now, you will have no difficulty in realising that the answer to our question is (b).

There is still one more aspect to cover. It concerns that category of 'weapons of offence' which become such only when they are carried with the intention of causing injury to the person.

Suppose Pickles was not armed with a gun during the burglary, but instead had in his pocket a knife which he intended to use to cause injury should the occasion arise. Piper, in whom he has confided, now knows of Pickles' possession of the knife and of his intention to use it. Collier knows only that Pickles has the knife but not of its intended use.

Remember this change of circumstances and complete the activity which follows.

Which of the following statements is/are correct?

(a) Pickles alone is guilty of aggravated burglary because he is the one who intends to use the knife offensively.

(b) Only Pickles and Piper are guilty of aggravated burglary because both know of the knife and knew of its intended use.

(c) All three are guilty of aggravated burglary because they all knew that Pickles had the knife with him.

(d) None is guilty of aggravated burglary because the knife is not a weapon of offence.

Answer: ....

In order to provide the correct answer to the question, it may be useful to break it down into essential components:

1. Is the knife a weapon of offence?

   Not in itself.

2. Is it to be used to cause injury to any person?

   Yes, if the occasion arises.

3. In that case, does it now become a weapon of offence?

   Yes.

4. Do they all know that Pickles 'has the knife with him'?

   Yes.

5. Do all three know of its intended use?

   Only Pickles and Piper.

6. Does Collier need to know of the intended use of the knife before being guilty of aggravated burglary?

   Yes.

Having asked those essential questions, and considered the answers to them, you will agree that only Pickles and Piper are guilty of aggravated burglary. The answer is (b).

'Has with him' therefore means knowingly having physical possession at the time of committing the burglary. In cases where two or more people commit burglary and one of them has physical possession of a prohibited article, then all who knew of the article are guilty of aggravated burglary.

When the article is a 'weapon of offence', which has become such only because the person 'having it with him' intends to use it to cause injury, then any co-defendant must, in addition to knowing of the article, also either know of the co-accused's intentions or have that intention themselves.

Summary

We have discussed in this section the basic ingredients of burglary and aggravated burglary and analysed the meaning of some of the words used in the legislation. If you want to have a more in-depth look at burglary you can consult the **Investigator's Manual**.

Self-assessment Test

Having completed this section, test yourself against the objectives outlined at the beginning of the section. You will find the answers below.

1. With regard to burglary under s. 9(1)(b), Theft Act 1968, assuming the person enters as a trespasser, in which of the following circumstances is the offence committed?
    (a) Pires climbed through an insecure house window at 4 am intending to rape the male occupier. The occupier was actually away on holiday so he poured paint over the furniture in the house.
    (b) Stone walked into the private staff area of a department store to use the telephone. On being discovered by a sales assistant Stone attacked the assistant with a steel chair attempting to cause grievous bodily harm.
    (c) Bennett broke into a shoe factory on Saturday evening. Once inside Bennett found no property worth stealing and started a small fire that set off the water sprinkler system causing £25,000 worth of damage.
    (d) Khan called at a house pretending to be water company official. The occupier invited Khan into the kitchen and was sent upstairs to turn on the bathroom taps. Khan followed and raped the occupier.

2. Fuller was part of a group drinking in a public house. Just before closing time Fuller went to the toilet. Once inside, Fuller decided to hide until everyone had left intending to sleep until the morning. With regard to burglary under s. 9(1)(a), Theft Act 1968, in which of the following circumstances is the offence committed?
    (a) After closing time, to see if the coast was clear, Fuller stood on the toilet pan to look over the cubicle door. Fuller slipped and broke the toilet pan causing £200 damage.
    (b) Fuller left the toilet at 2 am and went into the bar area to stretch out on a bench seat. Fuller saw the £50 bar float on a table and stole the money.
    (c) Fuller decided to leave with the money. Fuller realised the front door was alarmed so decided to break the toilet window to get out. Fuller went back into the toilet and used an elbow to break the toilet window.
    (d) Before leaving through the toilet window, Fuller decided to steal a framed photograph that was hanging on the toilet wall.

3. With regard to aggravated burglary under s. 10, Theft Act 1968, assuming that burglary is committed, in which of the following circumstances is aggravated burglary committed?
    (a) Dixon broke into a factory unit having entered through a ground floor window using a screwdriver. Dixon looked around for property to steal but left having found nothing worth stealing.
    (b) Wilkinson went into the private area of a leisure centre intending to steal a computer. Wilkinson had a plastic container with a strong corrosive liquid in it intending to use it to avoid leaving forensic evidence at the scene.
    (c) Tudor broke into a house and had a cigarette lighter that resembled an automatic pistol in his pocket. He stole a video recorder and left.

(d) McKay put a hand over the counter of a jewellers shop and stole a diamond ring from the display cabinet. McKay was wearing a wig and sun glasses as disguise.

4. Bryant and Dill went to a jeweller's home at 2 am to steal some valuable diamonds. When they entered the house, Bryant had a roll of strong adhesive tape with him and used it to tie the jeweller up. Dill knew Bryant had the adhesive tape and his intention from the outset but at no time touched it. However, Dill had not told Bryant that he had a toy gun in his pocket when they had entered the house which he intended to use to scare the jeweller if necessary.

   With regard to aggravated burglary under s. 10, Theft Act 1968, assuming that burglary is committed, which of the following statements is correct?
   (a) Both Bryant and Dill committed aggravated burglary because they both had the adhesive tape with them.
   (b) Only Bryant committed aggravated burglary with the adhesive tape because although Dill knew about the tape, he never had possession of it at any time.
   (c) Neither Bryant nor Dill committed aggravated burglary with the toy gun because it was not an imitation firearm for the purposes of the s. 10.
   (d) Bryant committed aggravated burglary with the gun because it is immaterial that he was unaware that Dill had the gun.

5. Mason and Floyd committed a burglary of a residential house in the middle of the afternoon. Floyd had taken a small kitchen knife with him for breaking into locked drawers in the house. Mason knew Floyd had the knife but Mason intended to use it to protect them from the occupier if necessary.

   With regards to aggravated burglary under s. 10, Theft Act 1968, which of the following statements is correct?
   (a) Both committed aggravated burglary. It does not matter that Floyd only had the knife for opening locked drawers, Mason had the required intent to cause injury.
   (b) Only Mason committed aggravated burglary because he had constructive possession of the knife and the required intent to use it as a weapon should the occasion have arisen.
   (c) Only Floyd committed aggravated burglary because he had the knife to facilitate the burglary.
   (d) Neither committed aggravated burglary because neither had the weapon with the necessary intent for the offence.

Answers to Self-assessment Test

1. Answer (b) because s. 9(1)(b) only includes attempted theft, theft, attempted grievous bodily harm or grievous bodily harm. Rape is not one of the ulterior offences for either section of burglary.

2. Answer (c) because Fuller intends to commit criminal damage on entering the toilet as a trespasser.

3. Answer (c) because the cigarette lighter has the appearance of being a firearm (there is no evidence that the screwdriver or fluid is intended to cause injury or incapacity).

4. Answer (a) because constructive possession is sufficient.

5. Answer (b) because the kitchen knife is not made or adapted to cause injury, a person can only commit the offence if they themselves have the required intent to use the knife to cause injury.

Fraud

Objectives

With regard to offences of fraud (ss. 15–17 and 24A, Theft Act 1968 and ss. 1 and 2, Theft Act 1978), at the end of this section, from a given set of circumstances, you will be able to:

1. Identify the points to prove for the offence of deception (s. 15, Theft Act 1968).

2. Distinguish between 'obtains' and 'appropriates'.

3. Recognise the different types of deception practised in an offence of deception.

4. Recognise *when* the deception must be practised.

5. Identify the points to prove for the offence of obtaining a pecuniary advantage by deception (s. 16, Theft Act 1968).

6. Identify the points to prove for the offence of false accounting (s. 17, Theft Act 1968).

7. Identify the points to prove for the offence of obtaining services by deception (s. 1, Theft Act 1978).

8. Identify the points to prove for the offence of evasion of a liability by deception (s. 2, Theft Act 1978).

9. Distinguish between 'remission', 'to wait' and 'exemption' in s. 2, Theft Act 1978.

10. Identify the points to prove for the offence of retaining a wrongful credit.

Introduction

Fraud is a word which makes the average investigator think of difficult and long enquiries. It paints a picture of many witness statements and mountains of documents. This of course is an extreme view of fraud, and makes up only a small percentage of the total number of frauds committed each year. The majority of frauds can be easily broken down into a series of simple thefts and deceptions with the occasional false accounting and forgery.

This section sets out to introduce you to the definition of deception and related topics.

Criminal Deception

Let us go back to Case Study 1 where Investigator Crouch is still completing the crime report. There were no direct clues at the jewellers to identify the suspect.

As luck would have it, Investigator Crouch is told that a suspect has just been detained at an off-licence using a stolen cheque book to obtain spirits. Investigator Crouch is given the task of interviewing the prisoner. Initial enquiries reveal that the cheque book used at the off-licence was stolen from 23 Elmwood Road, Sandford.

Investigator Crouch interviews the prisoner who makes certain admissions, as shown below in an extract from the tape recorded interview:

'... and I bought the cheque book and cheque card for £50 from a bloke called Billy, who I met in the Diamond pub. He told me he had got them from a house in Sandford. I altered the signature on the cheque card in my writing and I was trying to get some whisky on one of the cheques when you lot got me ...

... Last week I used one of the cheques to buy a leather coat in Marks and Spencer at the Keiller Shopping Centre ...'

 From all the information given to Investigator Crouch so far, how many offences should he be able to identify? List them below.

....

....

....

....

....

....

How did you get on? The offences are listed below with the Acts and sections to help you.

Offence	Section	Act
Criminal deception (re coat)	15	Theft Act 1968
Forgery of cheque (re coat)	1	Forgery and Counterfeiting Act 1981
Forgery of cheque card	1	Forgery and Counterfeiting Act 1981
Using forged instrument (re coat)	3	Forgery and Counterfeiting Act1981
Possession of forged instrument (cheques and cheque card)	5	Forgery and Counterfeiting Act 1981
Attempted criminal deception (re off-licence)	1	Criminal Attempts Act 1981
Attempted use of forged instrument (re off-licence)	1	Criminal Attempts Act 1981
Forgery (re off-licence)	1	Forgery and Counterfeiting Act 1981
Handling stolen goods	22	Theft Act 1968

In this Workbook we will be dealing with criminal deception. You will not be studying forgery for your examination.

Look now at the definition of criminal deception.

Section 15(1), Theft Act 1968

A person who by any deception dishonestly obtains property belonging to another, with the intention of permanently depriving the other of it, shall on conviction on indictment be liable to imprisonment ...

Having been reminded of this definition, indicate below how it differs from the definition of theft contrary to s. 1, Theft Act 1968.

Differences:

....

....

....

....

....

....

The differences are that for an offence under s. 15, Theft Act 1968, a deception is involved and the offender actually obtains the property.

Both theft and deception need the element of dishonesty. However, in deception the three provisions of s. 2 do not apply, so the jury or magistrate will apply the test in *Ghosh* to determine dishonesty. As in theft, there must be property belonging to another and an intention to permanently deprive.

The key differences between theft and deception are that in deception:

(a) the word 'obtains' is used instead of 'appropriates';

(b) deception must be practised and operating.

Why has the word 'obtains' been used in criminal deception? The reason is that 'obtains' has a wider meaning than 'appropriates' and tells us that the offender committing the deception can in fact get ownership of the property acquired, whereas a thief committing theft can never get ownership of the property stolen.

Practising a deception is not in itself an offence. Nor will an intention to deceive necessarily be proof of dishonesty.

Let us now look at what the word deception means.

Deception

Look at the definition of deception. It can be found in s. 15(4), Theft Act 1968.

Section 15(4), Theft Act 1968

... 'deception' means any deception (whether deliberate or reckless) by words or conduct as to fact or as to law, including a deception as to the present intentions of the person using the deception or any other person.

Let us go back to the deception admitted by the suspect in the interview. You remember the suspect used a stolen cheque book to obtain a leather coat from Marks and Spencer. Now complete the activity.

Below, the definition of deception has been broken down into its seven elements. You are given these in the first column of the table. Now from the information given to you about the obtaining of the leather coat, indicate which parts of the definition of deception apply in this case, by writing the reason why in the appropriate space.

Part of definition of deception	Does part apply? Yes/No	Give reason for your answer
Deliberate		
Reckless		
Words		
Conduct		
As to fact		
As to law		
Including a deception as to the present intentions of the person using the deception or any other person.		

Your table should look something like the one below, which continues on p. 40.

The deception to obtain the leather coat was not reckless nor did the suspect deceive the shop assistant as to law. It is true to say that most cases you will deal with will be deliberate deceptions and the victim will be deceived about some fact. The deception will probably involve both words and conduct together.

Interestingly, where a person remains silent and fails to act, that can amount to 'conduct' for the purpose of deception in some circumstances.

Having looked at what 'deception' means, there is a basic principle concerning when the deception comes in the order of events. Basically, for an offence of criminal deception, the deception must come before the obtaining of the property.

There are three points to prove in respect of the deception and the order of those points is important.

Part of definition of deception	Does part apply? Yes/No	Give reason for your answer
Deliberate	YES	The suspect knew that the cheque book was stolen and was deliberately deceiving the shop assistant into believing that the suspect was the owner of the cheque book.
Reckless	NO	Does not apply in this case because the deception was deliberate.
Words	YES	The suspect may have used words such as 'can I pay by cheque?', implying the cheque book was the suspect's and that there were funds in the bank.

Conduct	YES	The suspect's conduct alone could have deceived the shop assistant, because the normal way to pay for goods is to just take out a cheque book and pay by way of cheque without saying anything.
As to fact	YES	The suspect was deceiving the shop assistant as to the fact that the suspect was the owner of the cheque book etc.
As to law	NO	Does not apply in this case because the deception was as to fact and not to any legal requirement.
Including a deception as to the present intentions of the person using the deception or any other person.	YES	The suspect's intention was to obtain the leather coat and the intention at that time was to never pay for the coat.

Timing of Deception

1. A deception was practised.
2. The victim believed the deception.
3. The victim parted with property because of the deception.

You can also see that the deception must operate on the mind of the victim and he or she must part with the property as a result of the deception.

When a potential victim does not believe the deception, attempted deception would be the appropriate charge. Because the deception needs to operate on the human brain, it appears that a machine cannot be deceived.

When the deception is practised after the 'obtaining', no offence contrary to s. 15, Theft Act 1968, is committed and you would need to look to the Theft Act 1978 for possible alternative offences.

An additional section has been added to the Theft Act 1968 and deals with obtaining money transfer by deception from one account to another (usually a bank account or other business account) (s. 15A, Theft Act 1968). In this offence 'a person is guilty of an offence if by any deception he dishonestly obtains a money transfer for himself or another'.

A further related offence of retaining a wrongful credit has been created by s. 24A, Theft Act 1968. A wrongful credit is the credit side of a money transaction obtained contrary to any s. 15A offence or is derived from theft, blackmail or stolen goods. This section makes it an offence to keep a wrongful credit made to an account. It has the effect of ensuring that the proceeds from obtaining a money transfer by deception that is further transferred to another account will be deemed to be stolen goods for the purposes of 'handling'. You need to be aware of both of these offences but will find a full explanation in the **Investigator's Manual.**

Obtaining a Pecuniary Advantage

Section 16, Theft Act 1968

(1) A person who by any deception dishonestly obtains for himself or another any pecuniary advantage shall on conviction on indictment be liable to imprisonment ...

(2) The cases in which a pecuniary advantage within the meaning of this section is to be regarded as obtained for a person are cases where—

(a) [repealed]

(b) he is allowed to borrow by way of overdraft, or to take out any policy of insurance or annuity contract, or obtains an improvement of terms on which he is allowed to do so; or

(c) he is given the opportunity to earn remuneration or greater remuneration in an office or employment, or to win money by betting.

From this definition can you identify the difference between obtaining a pecuniary advantage and obtaining property by deception. Write your conclusions in the space below.

....

....

....

....

....

....

The answer may appear obvious but is still worthy of consideration. Obtaining a pecuniary advantage is where a person gets an opportunity by deceiving another person to gain a financial advantage. Obtaining property by deception is where a person uses a deception and actually gets some property as a result of that deception. You will notice that s. 16(2)(a) has been repealed. This section was replaced by the Theft Act 1978.

Obtaining Services by Deception

We have been looking at obtaining *property* by deception. However, people can obtain or evade other things by deception, and in the next part of this section we will cover some of these other offences.

Section 1(1), Theft Act 1978

A person who by any deception dishonestly obtains services from another shall be guilty of an offence.

Deception for obtaining services has the same meaning as defined in s. 15 of the 1968 Theft Act.

Similarly, with dishonesty, no guidance is given in the 1978 Act to the meaning of the word, so the appropriate points of s. 2 of the 1968 Theft Act will apply.

This leaves us with the obtaining of services and s. 1(2) defines it as 'an obtaining of services where the other is induced to confer a benefit by doing some act or causing, or permitting some act to be done on the understanding that the benefit has been, or will be paid for'. This definition sounds rather long-winded but relates to services where payment or the prospect of payment is required and excludes services which are done as favours such as lending your next-door neighbour a lawnmower.

Under s. 1, 'paid for' includes payment by money, cheque, credit card or in exchange for other goods.

Under s. 1, Theft Act 1978, a 'service' does not need to be lawful or contractual, so this would include the services of a prostitute or the placing of a bet. A 'service' includes the granting of a loan, such as from a bank.

Evading a Liability by Deception

Section 2(1), Theft Act 1978

... where a person by any deception—

(a) dishonestly secures the remission of the whole or part of any existing liability to make a payment, whether his own liability or another's; or

(b) with intent to make permanent default in whole or in part on any existing liability to make a payment, or with intent to let another do so, dishonestly induces the creditor or any person claiming payment on behalf of the creditor to wait for payment (whether or not the due date for payment is deferred) or to forgo payment; or

(c) dishonestly obtains any exemption from or abatement of liability to make payment; he shall be guilty of an offence.

 From the circumstances below, identify which offence has been committed contrary to s. 1 or 2, Theft Act 1978. Write your answer in the space provided in the right-hand column.

Circumstances	Offence committed
At the end of a meal in a restaurant Meikle received the bill and falsely told the waiter that she was very friendly with the owner of the restaurant. Meikle told the waiter that she would pay the owner the next time she saw her. By telling the waiter the name and address of the owner she convinced the waiter that she would settle the bill with the owner. Meikle only did this to avoid payment completely.	
Shah went to a tool hire shop. He falsely stated that he worked for Henderson Builders Ltd, a well-known local company who had an account with the shop. As a result Shah obtained a power drill for the day and then returned it without paying for it.	
Smithers was employed as a sales assistant and went to a leisure centre on his day off. When he went to pay he noticed that students were entitled to a 50% reduction on the entry fee. Smithers told the receptionist that he was a full-time student at the local college but had left his student card at home. As a result of this deception Smithers was given entry at the reduced rate.	
Stewart was unemployed and hired a car on a fixed mileage rate. When she returned the car the speedometer indicated that she had travelled 200 miles. After the hire costs had been calculated by the sales clerk, Stewart overheard another customer getting a 10% discount for being employed by a local company. Stewart told the sales clerk that she also worked for the local company and, in deceiving the sales clerk, received a 10% discount.	

Compare your answers with those in the table below.

Circumstances	Offence committed
At the end of a meal in a restaurant. Meikle received the bill and falsely told the waiter that she was very friendly with the owner of the restaurant. Meikle told the waiter that she would pay the owner the next time she saw her. By telling the waiter the name and address of the owner she convinced the waiter that she would settle the bill with the owner. Meikle only did this to avoid payment completely.	Meikle has, by using a deception, dishonestly induced the waiter (who is a person claiming payment on behalf of the creditor) to wait for payment, intending to make permanent default in payment. A liability exists and this is an offence contrary to s. 2(1)(b). Had Meikle written out a cheque on receipt of the bill (instead of telling the lies) intending that the cheque would never be honoured, this would induce the waiter to wait for payment. This is an offence contrary to s. 2(1)(b) also.
Shah went to a tool hire shop. He falsely stated that he worked for Henderson Builders Ltd, a well-known local company who had an account with the shop. As a result Shah obtained a power drill for the day and then returned it without paying for it.	Shah has, by using a deception, dishonestly obtained the use of the drill for the day. This is a 'service' and an offence contrary to s. 1. There is an overlap here and Shah also commits an offence contrary to s. 2(1)(c). Shah obtains an exemption from a liability to pay for the hire.
Smithers was employed as a sales assistant and went to a leisure centre on his day off. When he went to pay he noticed that students were entitled to a 50% reduction on the entry fee. Smithers told the receptionist that he was a full-time student at the local college but had left his student card at home. As a result of this deception Smithers was given entry at the reduced rate.	By paying a lower rate Smithers has obtained a reduction (abatement) of the liability to pay the full entry fee. The deception is used to reduce the liability before Smithers becomes liable for the entry fee. This is an offence contrary to s. 2(1)(c).
Stewart was unemployed and hired a car on a fixed mileage rate. When she returned the car the speedometer indicated that she had travelled 200 miles. After the hire costs had been calculated by the sales clerk, Stewart overheard another customer getting a 10% discount for being employed by a local company. Stewart told the sales clerk that she also worked for the local company and, in deceiving the sales clerk, received a 10% discount.	Stewart used a deception to get a reduction (remission) of part of an existing liability to make payment. This is contrary to s. 2(1)(a).

In contrast to s. 1, s. 2 requires a liability that is legally enforceable. This means that a bet or immoral service would not fall within s. 2. Section 2(1)(c) differs from s. 2(1)(a) and 2(1)(b) in that the liability does not need to be in existence at the time of the offence. Due to the nature of these offences there is likely to be an overlap between them and more than one offence may be committed in any one set of circumstances.

False Accounting

Section 17(1), Theft Act 1968

Where a person dishonestly, with a view to gain for himself or another or with intent to cause loss to another—

(a) destroys, defaces, conceals or falsifies any account or any record or document made or required for any accounting purpose; or

(b) in furnishing information for any purpose produces or makes use of any account, or any such record or document as aforesaid, which to his knowledge is or may be misleading, false or deceptive in a material particular;

he shall, on conviction on indictment, be liable to imprisonment.

This offence has two elements: altering an accounting document and using such an altered document. It is a supplementary offence to many of the fraud offences and is useful to you as an investigator because there is no need to prove an intention to permanently deprive.

Summary

You will see that fraud in general involves an element of deception. These offences will occur on a daily basis. However, more extensive frauds can become very complicated and involve company law. You will need to seek help with such investigations.

Self-assessment Test

Having completed this section, test yourself against the objectives outlined at the beginning of the section. You will find the answers below.

1. Frost found a bank cash-point card together with the PIN number. She used it in an automated cash machine to withdraw £100 cash. She had no bank account at that bank. With regard to the offence of obtaining property by deception (s. 15, Theft Act 1968), which of the following statements is correct?
   (a) Frost committed the offence contrary to s. 15 because 'obtains' means the same as 'appropriates'.
   (b) Frost did not commit the offence contrary to s. 15 because a machine cannot be deceived.
   (c) Frost did not commit the offence contrary to s. 15 because the deception operates at exactly the same time as the obtaining.
   (d) Frost committed the offence of deception because money was obtained from the machine.

2. Pettit had no authority to collect money. He visited a dwelling house and claimed he was collecting money for the Royal National Lifeboat Institute charity. The house owner quickly realised that Pettit was not representing the organisation and gave Pettit a £5 donation in order to trap him and then telephoned the police. With regard to the offence of obtaining property by deception (s. 15, Theft Act 1968), which of the following statements is correct?
   (a) Pettit deliberately used words and conduct to deceive the house owner about the fact that he was a bone fide collector.
   (b) Pettit committed the offence contrary to s. 15 notwithstanding that the house owner was not deceived.

(c) Pettit did not commit the offence contrary to s. 15 because any person can claim to collect money on behalf of a charity provided they give the money to charity.

(d) Pettit did not commit the offence contrary to s. 15 because there is no 'actus reus' present.

3. Jackson applied for a job at Norton Ltd as a structural engineer. At his interview he lied about his qualifications and previous experience. As a result of what Jackson said at the interview, he was given the job. One week later the company found out about the lies Jackson told and sacked him without pay. With regard to the offence of obtaining a pecuniary advantage (s. 16, Theft Act 1968), which of the following statements is correct?

(a) Jackson committed the full offence because he obtained the interview by deception.

(b) Jackson did not commit the offence contrary to s. 16 because he had not received his salary at the end of the first month's employment.

(c) Jackson committed the full offence because he secured an opportunity to earn money.

(d) Jackson did not commit the offence because the offence involves the obtaining of an improvement in the conditions of the employment.

4. Brooks was on a business trip and spent a night at a motorway hotel. He ran up a bill of £105 for bed, breakfast and an evening meal. He went to settle the bill and realised that his company did not have any agreement with the hotel about payment. Knowing that the hotel gave 20 per cent discount for certain customers, he deceived the receptionist to get his bill reduced by 20 per cent. With regard to evading a liability (s. 2(1)(a), Theft Act 1978), which of the following statements is correct?

(a) Brooks did not commit the offence contrary to s. 2(1)(a) because the £105 was an excessive amount for the services provided.

(b) Brooks did not commit the offence contrary to s. 2(1)(a) because the liability of £105 was not legally enforceable.

(c) Brooks committed the offence contrary to s. 2(1)(a) because the receptionist had a duty to confirm the reduction policy.

(d) Brooks committed the offence contrary to s. 2(1)(a) because the £105 was an existing liability.

5. Potter ordered a suit for £180 from a mail order catalogue. When it arrived, she decided she would keep the suit but put off paying for it by sending a cheque from her bank account. With regard to evading a liability (s. 2, Theft Act 1978), which of the following statements is correct?

(a) Potter committed the offence under s. 2(1)(b) of inducing the mail order catalogue to wait for payment, provided she knew the cheque would never be honoured.

(b) The £180 does not amount to an existing liability to pay the mail order catalogue.

(c) Potter committed the offence under s. 2(1)(b) of inducing the mail order catalogue to wait for payment, even though she intended to pay eventually.

(d) The suit amounts to an existing liability.

6. Bull went into a dentist and falsely stated that she was exempt from paying charges for dental treatment. The receptionist believed her and Bull had two teeth filled. With regard to ss. 1 and 2, Theft Act 1978, which of the following statements is correct?

(a) Bull only committed the offence of obtaining a service by deception contrary to s. 1.

(b) Bull only committed the offence of evading a liability by deception contrary to s. 2(1)(c).

(c) Bull committed both the offence of obtaining a service by deception contrary to s. 1 and evading a liability by deception contrary to s. 2(1)(c).

(d) Bull committed no offence because there needs to be an existing liability.

Answers to Self-assessment Test

1. Answer (b) because a machine cannot be deceived.

2. Answer (a) because his behaviour amounts to 'words or conduct'.

3. Answer (c) because this part of the offence involves the opportunity to earn money.

4. Answer (d) because the £105 represents an existing liability.

5. Answer (a) because there needs to be an intent to make permanent default.

6. Answer (c) because there is overlap between the offences.

Handling stolen goods

Objectives

With regard to offence of handling stolen goods (s. 22, Theft Act 1968), at the end of this section, from a given set of circumstances, you will be able to:

1. Identify the points to prove for the offence of handling stolen goods.
2. Recognise at what stage of the offence dishonesty must take place.
3. Recognise the offences by which 'goods' can be regarded as stolen.
4. Apply the provision for proving guilty knowledge or belief in handling stolen goods afforded by s. 27(3), Theft Act 1968.

Introduction

If you remember in Case Study 1, Investigator Crouch's suspect admitted buying a cheque book in a public house for £50. You might assume from the circumstances that the cheque book, being in a different name to his own or the seller's, was stolen. It looks likely that an offence of handling stolen goods has taken place. Have a look at the definition of the offence and try the first activity.

Handling Stolen Goods

Section 22(1), Theft Act 1968

A person handles stolen goods if (otherwise than in the course of the stealing) knowing or believing them to be stolen goods he dishonestly receives the goods, or dishonestly undertakes or assists in their retention, removal, disposal or realisation by or for the benefit of another person, or if he arranges to do so.

If you look close enough you should see two ways that the offence of handling stolen goods can be committed.

If the offence can be committed in two ways, try to identify each one below. Use the definition to assist you if you wish.

....

....

....

....

....

....

The offence can be broken down as follows:

A person handles stolen goods if (otherwise than in the course of the stealing) knowing or believing them to be stolen goods he dishonestly:

(a) receives them or arranges to receive them;

(b) assists or acts for the benefit of another person.

There are several points within the definition but here are the basic aspects:

Dishonesty

Dishonesty has the same meaning as in *R* v *Ghosh*. The dishonesty must be at the time of handling.

Atwal v Massey [1971] 3 All ER 881

'The dishonesty must be at the time of the relevant act of handling.'

Where a person takes possession of goods but does not know or believe that the goods are stolen, they are not guilty of handling. However, they would be guilty of theft if they had not paid value for the goods, and then dishonestly kept the goods on finding out that they were stolen.

Goods

In handling stolen goods, the word 'goods' has been used instead of the word 'property'. With a few exceptions they mean basically the same thing.

Look at the definition of 'goods' as given by s. 34(2)(b), Theft Act 1968.

Section 34(2)(b), Theft Act 1968

'goods' except in so far as the context otherwise requires, includes money and every other description of property except land, and includes things severed from the land by stealing.

You can see that the definition of 'goods' states that it includes 'every description of property'. Therefore, this phrase shows us that real, personal, things in action and other intangible property also come within the definition.

Stolen

Lastly, if goods are not stolen there is no handling.

Let us look at when, for the purpose of this offence, goods are regarded as being stolen.

There are three types of offences where goods are considered to be stolen within the meaning 'handling stolen goods'. They are listed below.

Under s. 24(4), Theft Act 1968, goods obtained in England and Wales or elsewhere either by:

(a) theft (s. 1, Theft Act 1968);

(b) blackmail (s. 21, Theft Act 1968);

(c) the circumstances described in s. 15(1), Theft Act 1968.

In addition, the definition of goods includes the subject of an act committed in a foreign country which was a crime by the law of that country and which, had it been committed in England or Wales, would have been theft, blackmail or obtaining by deception.

Note that goods obtained outside England and Wales can still be 'handled' here. You can also see that s. 15 is not referred to simply as 'criminal deception'.

Goods are no longer considered stolen once they have been restored to:

(a) the person from whom they were stolen;

(b) another person's lawful possession or custody (e.g. police): or

(c) after that person (or any person claiming through him) has ceased to have any right or restitution in respect of the theft.

'Stolen Goods' will also include, in addition to the goods originally stolen and parts of them, any other goods, which directly or indirectly represent or have represented, at any time the stolen goods. This means the proceeds from stolen goods fall within the disposal or realisation referred to in this section. It is immaterial whether this disposal or realisation has been done by the person stealing the goods, or the subsequent handler.

Knowing or Believing

'Knowing or believing' is part of the *'mens rea'* for the offence and will often be a difficult element of the offence to prove. It is for this reason that s. 27(3) was included in the legislation.

Section 27(3), Theft Act 1968

Where a person is being proceeded against for handling stolen goods (but not for any offence other than handling stolen goods), then at any stage of the proceedings, if evidence has been given of his having or arranging to have in his possession the goods subject of the charge, or of his undertaking or assisting in, or arranging to undertake or assist in, their retention, removal, disposal or realisation, the following evidence shall be admissible for the purpose of proving that he knew or believed the goods to be stolen goods—

(a) evidence that he has had in his possession, or has undertaken or assisted in the retention, removal, disposal or realisation of, stolen goods from any theft taking place not earlier than twelve months before the offence charged; and

(b) (provided that seven days' notice in writing has been given to him of the intention to prove the conviction) evidence that he has within the five years preceding the date of the offence charged been convicted of theft or of handling stolen goods.

This means that the court can take previous character into account when deciding guilty knowledge. However, this section is used sparingly in practice. Therefore, as an investigator, you will need to investigate all the surrounding circumstances to prove whether a suspect 'knew or believed' that the goods were stolen.

Mere suspicion by a suspect will not be sufficient to prove 'knowing or believing'. Nor will the suspect turning a blind eye in itself be enough. However, these two things with other aspects of the circumstances could, as a whole, prove 'knowing or believing'. These other aspects of the circumstances can be remembered with the use of the mnemonic HANDLINGS.

What do the letters HANDLINGS in the mnemonic stand for?

H ....

A ....

N ....

D ....

L ....

I ....

N ....

G ....

S ....

You should have identified the following aspects of the handling that would go towards proving 'knowing or believing'. These aspects add weight to the evidence of 'knowing or believing'.

H: Handling the goods.

A: Accepting goods from a known thief.

N: No receipts.

D: Denial of possession and subsequently found in possession.

L: Low price paid for the goods.

I: Identification marks removed.

N: No explanation offered.

G: Goods received at odd times in unusual circumstances.

S: Several times having received goods from the same person.

Summary

Handling stolen goods has many facets to it, but for practical purposes the offence can be dealt with as two parts. In collecting evidence you will need to prove 'knowing or believing' and this will often mean investigating all the surrounding circumstances of the handling.

Self-assessment Test

Having completed this section, test yourself against the objectives outlined at the beginning of the section. You will find the answers below.

1. Curson decided to break into a bonded warehouse to steal a lorry load of spirits. A month before the burglary he telephoned French who was a wholesaler always ready to buy stolen spirits. Curson and French agreed that as soon as the spirits were stolen they should be taken immediately to a secret warehouse nearby. With regard to the offence of handling stolen goods (s. 22, Theft Act 1968), which of the following statements is correct?
   (a) French cannot be guilty of handling stolen goods because no property has been stolen.
   (b) French is guilty of handling stolen goods as soon as he agrees to receive the spirits.
   (c) The spirits can never be stolen property for the purpose of handling because burglary is not one of the three offences mentioned in s. 24(4), Theft Act 1968.
   (d) Evidence of a telephone conversation is not admissible in an offence of handling stolen goods.

2. With regard to the offence of handling stolen goods (s. 22, Theft Act 1968), which of the following statements is correct?
   (a) Moore bought a mobile phone very cheaply from a colleague in her office at work. When she bought the phone she thought it was work surplus and being scrapped. Later she was told that a similar mobile phone had been stolen from another department at work. She made some enquiries and confirmed that the phone she had bought was in

fact stolen. She decided that she would keep the phone and made sure that she did not take it into work. Moore can never be guilty of handling under these circumstances.

(b) Massey was charged with receiving stolen cigarettes. He denied knowing or believing that the cigarettes were stolen. The fact that he had a previous conviction for theft within the five years preceding the date of the offence charged would be admissible to prove his dishonesty under s. 27(3) Theft Act 1968.

(c) Stevens buys some land as a building plot knowing it to be stolen property. He is guilty of handling stolen goods as soon as he enters into an agreement to buy the land.

(d) Property cannot be stolen goods for the purpose of handling when it is obtained through blackmail because a loss may be involved.

Answers to Self-assessment Test

1. Answer (a) because stolen goods must be in existence.

2. Answer (a) because Moore must 'know or believe' at the time of the receiving. Massey's conviction has to be for theft or handling stolen goods and it goes to prove 'knowing or believing' not dishonesty. Land is excluded. Blackmail is an offence under s. 24(4).

PART TWO

Assaults, Drugs, Firearms and Defences

Assault

Objectives

With regard to the offences of assault (Offences Against the Person Act 1861), at the end of this section, from a given set of circumstances, you will be able to:

1. Identify an assault and a battery.

2. Apply the defences to a charge of assault.

3. Identify the points to prove for 'grievous bodily harm/wounding' (ss. 18 and 20, Offences Against the Person Act 1861).

4. Identify the points to prove for assault with intent to resist arrest (ss. 38, Offences Against the Person Act 1861).

5. Identify racially or religiously aggravated assaults (s. 29, Crime and Disorder Act 1998).

6. Identify the points to prove for assault on police (s. 89, Police Act 1996).

7. Identify the points to prove for threats to kill (s. 16, Offences Against the Person Act 1861).

8. Identify the offences of false imprisonment, kidnap and child abduction.

Introduction

In this section we will be dealing with assault.

In the space below write down any definition of assault that you know:

....

....

....

....

....

....

There is no right answer to this question, but there is one definition regularly used. Have a look at it below.

> Assault includes the intentional or reckless application of unlawful force to the person of another without his consent, or the threat of such force by act or gesture, if the person threatening has caused, or causes the person threatened, to believe that he has the present ability to effect his purpose.

An assault is a battery if the force is actually applied and is a wounding if the flesh is opened.

Another definition sometimes put forward is very similar, this is:

> An assault is an attempt, offer or threat to use any unlawful force to another. It is constituted by any attempt to apply unlawful force to another, or any threat which is accompanied by or consists of any act or gesture showing a present intent to use unlawful force and is also accompanied by a present ability to carry the threat into execution.

You should make yourself familiar with the definition of assault.

Generally, there are few problems with concluding that an assault has occurred when force is actually applied and the degree of injury will decide the course of action to be taken.

'Reckless' is referred to in the definition and means recklessness as to whether such harm should occur or not (i.e. the accused has foreseen that the particular kind of harm might be done, and yet has gone on to take the risk of it). It is neither limited to, nor does it require, any ill will towards the person who was in fact injured.

It has been generally accepted that words alone do not constitute an assault. Previously, there had to be some physical movement in the form of some act or gesture which caused the victim to apprehend that they were about to be struck.

However, in *R v Ireland* [1998] AC 147, the question of psychological injury was raised.

In this case the defendant made a large number of unwanted telephone calls, between June and September 1994, to three women, frequently over a short period. When the calls were answered, the defendant was silent. Each woman complained of significant psychological symptoms, such as palpitations, difficulty in breathing, cold sweats and tearful disposition brought upon by nervousness. In considering the defendant's appeal against conviction, the Court of Appeal held:

1. An assault was an act whereby a person intentionally or recklessly caused another to apprehend immediate and unlawful violence.

2. Whether a particular act amounted to an assault was a question of fact which would depend on all the circumstances of the case.

3. The making of telephone calls followed by silence, or a series of such calls, was an act capable of constituting an assault under s. 47, Offences Against the Person Act 1861. The act consisted of making the call and it was irrelevant whether words or silence ensued.

The calls made the victims apprehensive and caused psychological damage.

A threat to inflict harm at some time in the future cannot amount to an assault, an apprehension of immediate personal violence is essential. The accused must intend to cause the victim to believe that he or she can and will carry the threat out and the victim must believe that he or she will.

Where there is an application of force (i.e. a battery), there is no need for there to be an apprehension of impending violence. For example, a blow from behind of which the victim is totally unaware will still be assault.

The merest touching, without consent, is legally a criminal offence. However, there is implied consent to all physical contact which is generally acceptable in the ordinary conduct of daily life, e.g. jostling in a supermarket, at a football match, on a crowded bus etc.

It is important to understand these basic principles of assault, as they are the basis of the more serious assaults which you will have to investigate.

Defences

When is an assault not an offence?

Give as many circumstances as you can in which the use of force, or threatened use of force, is not an offence.

....

....

....

....

....

....

The defences are:

(a) Accident.

(b) Consent given freely by a rational and sober person knowing the nature of the act. If under 16 years cannot consent to any assault.

(c) Moderate correction of a child or young person by parent of guardian.

(d) Lawful sport if within the rules of the game and not an unreasonable and unjustifiable amount of physical force used.

(e) Self-defence.

(f) Defence of your family.

(g) Execution of a legal or official duty.

(h) Defence of your property.

Defences (e)–(h) only apply providing in each case that no more force is used than is reasonable in the circumstances. What is 'reasonable' will be for a court to decide.

These defences come from the common law and s. 3, Criminal Law Act 1967.

Let us now look at the specific offences of assault. We will do this through the eyes of the victim.

Case Study 4

An extremist nationalist political party decided to organise a demonstration against immigration into the country by marching through the centre of a town with a high percentage of ethnic minority groups. The chief constable of the area, fearing that this would cause a serious threat to public order, successfully applied to have the march banned. The leaders of the extremists decided to make their point despite the ban. Unknown to the local police, on a Saturday lunchtime they arranged an informal meeting in a large pub in the town centre. After several drinks, two of group, Gary and Steve, went out into the street to deliberately provoke trouble. Steve stabbed an Asian youth with a broken beer bottle causing a ten centimetre gash in the groin. The injury to the youth needed blood transfusions and extensive surgery. Gary pushed a second Asian youth hard in the back causing him to fall over and stamping on his head causing extensive bruising to his face. A member of the public rushed forward and told Gary and Steve to stop. Gary pushed her aside and said 'Keep your nose out of it'. She fell over and suffered minor bruising. After the assaults, Gary and Steve left their victims on the pavement and ran off. A witness to the assaults chased the two men to detain them. The witness caught Steve and grabbed him around the neck in a head-lock. In order to escape Steve kneed the witness in the stomach and got away. The witness was winded but not injured.

From these circumstances, identify the type of assault inflicted on each victim.

Victim	Type of assault	Act and section
Asian youth		
Second Asian youth		
Member of the public		
Witness		

Your table should have looked like this:

Victim	Type of assault	Act and section
Asian youth	GBH/Wounding	Section 18 or 20, Offences Against the Person Act 1861.
	Racially aggravated	Section 29(1)(a), Crime and Disorder Act 1998.
Second Asian youth	ABH	Section 47, Offences Against the Person Act 1861.
	Racially aggravated	Section 29(1)(b), Crime and Disorder Act 1998.
Member of the public	Common Assault/Battery	Common Law and s. 39, Criminal Justice Act 1988.
	Racially aggravated	Section 29(1)(c), Crime and Disorder Act 1998.
Witness	Assault with intent to resist lawful arrest.	Section 38, Offences Against the Person Act 1861.
	Common Assault/Battery	Common Law and s. 39, Criminal Justice Act 1988.

Taking the most serious assault first, from the circumstances described, it is not possible to determine whether a s. 18 or s. 20 assault has been committed on the Asian youth. However, an important aspect of these circumstances is the racially aggravated nature of the assaults. Where a magistrate or jury are satisfied that an assault is racially aggravated, the aggravated form of the offence attracts a longer prison sentence. This would also be the case for religiously aggravated assaults. The maximum penalty for a common assault increases from six months to two years' imprisonment, actual bodily harm or s. 20 assault increases from five to seven years' imprisonment. There is not a similar increase for a s. 18 assault because it already has a maximum penalty of life imprisonment. However, where a s. 18 offence is racially aggravated, the judge will take this into account when sentencing. It is for these reasons that you must ensure that where an assault is racially motivated, this aggravating factor is thoroughly investigated.

You will be aware that the offence committed depends on the seriousness of the injury caused.

The Crown Prosecution Service use a table of 'Charging Standards in Cases of Assault'. This table is shown below and may provoke some argument, but remember there will always be room for interpretation considering the surrounding circumstances.

	CHARGE			
	Common assault	ABH s. 47	GBH s. 20	Wounding s.18
Graze	*			
Scratch	*			
Abrasion	*			
Minor bruising	*			
Swelling	*			
Reddening to skin	*			
Superficial cuts	*			
Black eye	*			
Loss of/broken tooth		*		
Temporary loss of sensory function (includes consciousness)		*		
Extensive/multiple bruising		*		
Displaced/broken nose		*		
Minor fracture		*		
Minor (not superficial) cuts		*		
Permanent disability			*	*+
Permanent loss of sensory function			*	*+
Visible disfigurement (not minor)			*	*+
Broken/displaced limbs or bones			*	*+
Injury causing substantial blood loss			*	*+
Injury resulting in lengthy treatment/incapacity			*	*+

*+ **Section 18** requires an intent to wound or cause grievous bodily harm and infliction of harm.

Evidence of intent might include a repeated or planned attack, deliberate selection of a weapon, adaptation of an article to cause injury, prior threats, use of an offensive weapon or kicking a victim's head.

Assault on constable in execution of duty Section 51(1), Police Act 1964		Assault with intent to resist arrest Section 38, Offences Against the Person Act 1861
If evdience of 'execution of duty' is not available, consider s. 39, Criminal Justice Act 1988 (common assault), or s. 47, Offences Against the Person Act 1861.	⟶	Can be used for assaults on members of the public, e.g. store detectives. Unlike s. 51(1) of the Police Act 1964, this offence is indictable and can therefore be coupled with other indictable offences.

We will now look at the definition of each type of assault mentioned in the answer.

Grievous Bodily Harm/Unlawful Wounding

You will be called upon to investigate assaults under s. 18 and s. 20 of the Offences Against the Person Act 1861.

What is the difference between the two sections? From your experience, outline the differences below.

....

....

....

....

....

....

Have a look at the two definitions. You will see that the difference should be obvious.

Section 18, Offences Against the Person Act 1861

Whosoever shall unlawfully and maliciously by any means whatsoever wound or cause any grievous bodily harm to any person with intent to do some grievous bodily harm to any person, or with intent to resist or prevent the lawful apprehension or detainer of any person ...

Section 20, Offences Against the Person Act 1861

Whosoever shall unlawfully and maliciously wound or inflict any grievous bodily harm upon any other person ...

Clearly for an offence of s. 18, the offender must have an intent to commit grievous bodily harm. It also applies to where he or she causes grievous bodily harm or wounding with intent to resist his or her or another's arrest. This intent makes s. 18 a more serious assault carrying a greater custodial penalty than s. 20.

With a s. 20 assault, the offence merely requires an unlawful and malicious act resulting in either a wounding or grievous bodily harm.

Both sections contain the word 'malicious', which has been held to mean 'recklessly'. This means that the defendant must realise that there is a risk of some harm being caused to the victim. The defendant does not need to foresee the degree of harm that will be caused, only that their behaviour may bring about some harm to the victim.

It would appear that 'maliciously' only has a real effect in relation to s. 20 and the second part of s. 18. The first part of s. 18 requires a specific intent to wound or cause grievous bodily harm, i.e. it was the accused's purpose to cause grievous bodily harm, or if not, he or she must know that the grievous bodily harm was a virtual certain consequence of his or her act.

Section 18 also gives an alternative 'specific intent', i.e. to resist the lawful apprehension or detention of any person. The accused need *not* 'intend' to wound or cause grievous bodily harm in these circumstances but could be 'reckless' as to the outcome of his or her actions which, if carried out with the intention of resisting the lawful apprehension of any person, turns s. 20 circumstances into a s. 18 offence. So 'maliciously' does have an effect on the second part of s. 18.

The word 'inflict' in s. 20 has caused problems in the past. It is now clear that it is interpreted similarly to 'cause', in that an assault is not needed and harm can be 'inflicted' indirectly. An example would be menacing telephone calls that inflict psychiatric harm. It should be enough to show that the defendant's behaviour brought about the resulting harm.

The basic difference between the two offences is the intent behind the assault.

Thus, the difference between ss. 18 and 20 is that s. 18 requires an intent to do either of the following:

(a) some grievous bodily harm;

(b) resist the lawful apprehension or detention of any person.

Before we deal with s. 47, let us quickly look at the definitions of wounding and grievous bodily harm.

Wounding is defined in *Moriarty* v *Brooks* (1834) as:

> The human body has seven layers of skin and all layers must be broken to constitute a wounding. A mere abrasion is not enough but the splitting of the skin within the mouth is sufficient.

Grievous bodily harm is defined in *DPP* v *Smith* [1916] AC 290 as:

> Means really serious bodily harm. It is not necessary that the harm should be permanent.

Actual Bodily Harm

Let us now deal with the assault on the second Asian youth. Remember he received extensive bruising to the face.

Has he been wounded? No. Does his injury amount to grievous bodily harm? No. His injury is actual bodily harm.

How would you define the injury of actual bodily harm? Write your definition below.

....

....

....

....

....

....

Have a look at what 'actual bodily harm' means and compare it with your answer.

Actual bodily harm can be defined as:

An injury more than merely trivial and includes a hurt or injury which is calculated to interfere with the health or comfort of the victim.

That is the nature of the injury. The offence requires an *assault* plus *actual bodily harm*.

The offence is 'assault occasioning actual bodily harm' (s. 47, Offences Against the Person Act 1861).

There is not really a great deal more to be said about this offence. All that must be proved is that an assault took place inflicting injuries amounting to actual bodily harm.

As the punishment is exactly the same as for s. 20, it is not surprising that when an assault is present, prosecutors often opt to charge s. 47 avoiding the burden of providing a greater degree of harm.

Assaults with Intent to Resist Arrest

Section 38, Offences Against the Person Act 1861

Whosoever … shall assault any person with intent to resist or prevent the lawful apprehension or detainer of himself or of any other person for any offence …

This may be a useful provision where civilians attempting to make a 'citizen's arrest' are assaulted; however police officers, and those assisting them in the 'execution of their duty', are covered by the Police Act 1996.

Section 89, Police Act 1996

(1) Any person who assaults a constable in the execution of his duty, or a person assisting a constable in the execution of his duty, shall be guilty of an offence …;

(2) Any person who resists or wilfully obstructs a constable in the execution of his duty, or a person assisting a constable in the execution of his duty shall be guilty of an offence …

The main point of contention is usually: was the constable acting in the 'execution of his duty'? Each case will be decided on the facts, but generally, if the officer's conduct falls within '*his duty*' to prevent crime or apprehend offenders, he will have the protection of the law. Only if officers exceed their power will problems arise. For example, in *McArdle v Wallace* [1964] Crim LR 467, a constable made enquiries at a café regarding some stolen property. He was asked to leave and then assaulted. He was acting in the execution of his duty in making the enquiries, but once asked to leave he became a trespasser and was no longer acting in the 'execution of his duty'.

Usually assaults under this section will be common assaults but injuries to officers of a more serious nature can still be prosecuted under s. 89, Police Act 1996, although depending on the circumstances, charges of s. 47, etc. Offences Against the Person Act 1861 may be more appropriate.

Persons who assist police officers and are assaulted are also only protected by s. 89, Police Act 1996, if the officer is acting in the execution of his duty.

Threats to Kill

Section 16, Offences Against the Person Act 1861
(as amended by s. 12, Criminal Law Act 1977)

A person who without lawful excuse makes to another a threat, intending that the other would fear it would be carried out, to kill that other or a third person, shall be guilty of an offence ...

The points to remember are:

The prosecution must prove that the **offender** intended that the person to whom the threat is made would fear the threat would be carried out; it does not matter that the person receiving the threat is not concerned or even believed it.

The offence is complete as soon as the threat is made as long as the intent of the accused is present.

No killing is necessary.

Mode of Trial and Maximum Punishments

Offence	Mode of trial	Punishment
Section 18, Offences Against the Person Act 1861	Indictment only	Life imprisonment
Section 20, Offences Against the Person Act 1861	Either way	5 years' imprisonment on indictment (7 years if racially or religiously aggravated)
Section 47, Offences Against the Person Act 1861	Either way	5 years' imprisonment on indictment (7 years if racially or religiously aggravated)
Section 38, Offences Against the Person Act 1861	Either way	2 years' imprisonment on indictment
Assault police, Section 89, Police Act 1996	Summary only	6 months' imprisonment and/or fine
Obstruct police, Section 89, Police Act 1996	Summary only	1 month's imprisonment on indictment
Common assault, Common Law and s. 39, Criminal Justice Act 1988	Summary	6 months' imprisonment and/or fine (2 years if racially or religiously aggravated)
Threat to kill, Section 16, Offences Against the Person Act 1861	Either way	10 years' imprisonment on indictment

Other Offences

There are three further associated offences that involve an element of restraint or force. These are false imprisonment, kidnapping and child abduction. The first two are common law offences, while child abduction is contrary to the Child Abduction Act 1984. They are all arrestable offences.

False Imprisonment

It is an offence to falsely imprison another person. This effectively means the unlawful and intentional or reckless restraint of a person's freedom of movement at any place for any length of time.

Kidnapping

This is the more serious offence of intentionally or recklessly taking or carrying away another person without their consent and without lawful excuse. This amounts to more than mere restraint and can be effected by force or fraud.

Child Abduction

There are two offences of abducting children under the age of 16 contrary to the Child Abduction Act 1984. One offence concerns people who are 'connected with the child' and the other offence where the offender is not 'connected with the child'.

Section 1 makes it an offence for a person connected with a child to take or send the child out of the United Kingdom without appropriate consent. A connected person is normally a parent or person acting as the parent.

Section 2 is an offence similar to false imprisonment or kidnapping, but the victim is a child. The child may be consenting to the restraint and therefore negating the offences of false imprisonment and kidnapping. The offender is not a parent or person acting as a parent. The child does not need to leave the United Kingdom for the offence to be made out.

Summary

The seriousness of the assault quite naturally depends upon the severity of the injury caused by that assault and the intentions behind such an assault.

Self-assessment Test

Having completed this section, test yourself against the objectives outlined at the beginning of the section. You will find the answers below.

1. In which of the following circumstances is there *neither* an assault *or* a battery?
    (a) Teresa picked up a vase of flowers and smashed Will over the head with them.
    (b) Jim had an argument with Tom about some money Tom owed him. During the argument, Jim said 'Unless you pay me by next week you'll need plastic surgery'.
    (c) Tina was walking by the edge of the swimming pool when a youth deliberately pushed her into the deep end.

(d) James saw Mark from a distance in a crowded pub. There had been trouble between them in the past and James gestured with his fist that he would punch Mark.

2. In which of the following circumstances would a defence to an assault **not** be applicable?
   (a) A firm but fair tackle on Adam during a football match breaks his ankle.
   (b) Jane smacks her five-year-old son on the leg for climbing on to a kitchen worktop near to a boiling saucepan.
   (c) Henry fires a shotgun at some boys stealing apples from a tree in his garden.
   (d) Steven punches Carol in the face as she approaches him with a kitchen knife intent on injuring him.

3. In which of the following circumstances is the offence of wounding or causing grievous bodily harm with intent contrary to s. 18, Offences Against the Person Act 1861 made out?
   (a) Mike armed himself with a length of heavy gas piping intending to seriously injure Steve. He hit Steve over the head several times causing extensive bruising to the skull.
   (b) David was with some friends in a cinema. He was bored with the film and shouted 'fire! fire!' to annoy his friends. As a consequence, several people in the cinema panicked and ran for the exits. One elderly woman tripped and broke her leg.
   (c) Matt saw his friend being arrested by a police constable for being disorderly on a train platform. He approached the two and even though Matt realised it was dangerous so near the railway line, he pulled his friend away from the officer so his friend could escape. In doing so, the officer broke his arm in two places as he fell onto the railway line.
   (d) Chris had an argument with her best friend and, out of frustration, lashed out at her with a dinner plate. Chris had not intended to injure her friend but realised the danger of such action and did it anyway. Her friend received extensive wounds to the inside of her mouth.

4. Where an offence is racially aggravated (s. 29, Crime and Disorder Act 1998), which of the following statements is *incorrect*?
   (a) The maximum penalty for common assault increases to two years' imprisonment.
   (b) The maximum penalty for actual bodily harm increases to seven years' imprisonment.
   (c) The maximum penalty for s. 20, Offences Against the Person Act 1861 increases to ten years' imprisonment.
   (d) There is no increase in the maximum penalty of life imprisonment for s. 18, Offences Against the Person Act 1861.

5. Which of the following statements is *incorrect*?
   (a) Dave, a science teacher, commits an offence contrary to s. 2, Child Abduction Act 1984 when he takes one of his 15-year-old male pupils on a week's holiday in Cornwall without the consent of the pupil's parents. The pupil is a willing party to the holiday.
   (b) Anne commits the common law offence of false imprisonment when she parks her car in the street, activates the central locking and refuses to let her boyfriend out of the car for five minutes until he agrees to buy a new car.
   (c) Mark commits the common law offence of kidnapping when he will not allow a hitch-hiker out of his car when she states she wants to leave. He drives off with her looking for a place to sexually assault her.
   (d) Kevin is separated from his wife who has custody of their six-year-old daughter in England. Kevin commits the offence contrary to s. 1, Child Abduction Act when he takes his daughter to Wales and keeps her there without the consent of his wife.

Answers to Self-assessment Test

1. Answer (b). A threat to inflict harm at some time in the future cannot amount to an assault.

2. Answer (c). The defence of property only applies providing the force used is reasonable in the circumstances.

3. Answer (c). There is specific intent to prevent arrest even though Matt is merely reckless as to any injury being caused. A broken arm is a serious injury.

4. Answer (c). The maximum penalty increases to seven years' imprisonment.

5. Answer (d). Section 1 deals with a person connected with the child taking that person out of the UK without the appropriate consent.

Homicide

Objectives

With regard to homicide, at the end of this section, from a given set of circumstances, you will be able to:

1. Identify the points to prove for the offence of murder (Common Law).
2. Identify the points to prove for the offence of solicitation of murder (s. 4, Offences Against the Person Act 1861).
3. Identify the special defences to murder (ss. 2–4, Homicide Act 1957).
4. Identify the points to prove for the offence of involuntary manslaughter (Common Law).
5. Distinguish between court finding of voluntary manslaughter (ss. 2–4, Homicide Act 1957) and the offence of involuntary manslaughter (Common Law).
6. Identify the points to prove for the offence of aiding suicide (s. 2, Suicide Act 1961).
7. Identify the difficulty in proving the offence of corporate manslaughter (Common Law).

Introduction

As you know, homicide is a general term used for offences that bring about the death of a person. Although murder is usually considered the most serious offence that can be committed, the law itself is not particularly complicated for you as an investigator. However, it can get very complicated for the court. Involuntary manslaughter can be a little more involved but this section will help you distinguish between voluntary and involuntary manslaughter. In this section you are given a comprehensive series of case scenarios to apply your understanding of the law.

Murder

Look at the common law offence of murder below and then work through the case scenarios that follow.

Common Law

Murder is committed when a person unlawfully kills another human being under the Queen's Peace, with malice aforethought.

Now decide from the facts provided in the scenarios whether you think the suspect(s) identified in each case has committed murder. Give reasons for your answers. The answer is provided immediately after each activity.

Activity

Case scenario	Decision and reasons
Liam had a very stormy relationship with his partner Katie. She became pregnant as a result of a brief affair with another man. Liam was not happy that Katie was intent on keeping the baby. Towards the end of the pregnancy, Liam became increasingly violent towards Katie. One night he came back to the house and attacked Katie with a weapon intending to kill the unborn baby. Katie was rushed to hospital where the baby was delivered alive but died two years later as a result of the injuries received in the original attack. Katie made a full recovery.	

'Another human being' and 'year and a day'

From the definition of murder you can see that a human being has to be killed. Whilst the baby is still in the womb it is not, in law, a human being and cannot be murdered (the offence of child destruction, Infant Life (Preservation) Act 1929, could apply when the child is still in the womb). However, once the baby has been born, and has had an independent existence from the mother, it is a human being. Therefore, in this case Liam committed the offence of murder of the baby. The baby had an independent existence for two years outside of the womb. He intended to kill the baby and his attack caused the death of a human being. In slightly different circumstances, Liam may have intended only to seriously injure Katie. In this latter case, had the baby died after birth, there can be no transfer malice from Katie to the baby and therefore no murder of the baby (later you will see that involuntary manslaughter of the baby could be committed in these circumstances).

There used to be a period (one year and a day) during which time the victim of the murder had to die for it to be murder. That rule has been abolished because advances in medical science can now keep people alive in a vegetative state for long periods of time. However, where the victim dies more than three years after they received their injuries, the consent of the Attorney-General is needed in order to proceed.

Activity

Case scenario	Decision and reasons
Police Constable Fry was a member of a firearms team. He was deployed as part of an operation to arrest a team of armed robbers for conspiracy to rob a post office. The operation plan was to arrest the men as they left their car and before they entered the post office. He was fully briefed and complied with all the necessary regulations regarding firearms incidents. When the robbers' car pulled up outside the post office the firearms team moved in. Police Constable Fry shouted the appropriate warnings and directions to one of the robbers. The robber did not comply with the warnings and directions. The robber raised the shotgun he was carrying and pointed it directly at Police Constable Fry. Police Constable Fry shot the robber in the chest and killed him.	

'Unlawful'

You can see from the definition of murder that the killing has to be unlawful. Most killings where the offender intends to kill will usually be unlawful. However, the State has a duty to protect the life of its citizens and police officers will on rare occasions, and as a last resort to protect life, be authorised to kill a human being. In the circumstances described, it would appear that Police Constable Fry was acting fully within the strict parameters set for the use of lethal force. He does not commit the offence of murder or manslaughter. It would also be lawful for a person to use reasonable force in self-defence.

Activity

Case scenario	Decision and reasons
Graham was a car dealer who was in the habit of making enemies. He made the mistake of double-crossing another car dealer called Sam. Sam was not prepared to be messed about and had some violent friends in the criminal underworld. Sam decided that Graham should be killed. Sam met one of his boyfriends and agreed to pay him £10,000 to kill Graham. After some research and preparation, the boyfriend went to Graham's holiday address in southern Spain early one morning. He followed him when he took his dog for his usual morning exercise. When Graham got to an isolated coastal path the boyfriend moved quickly up behind Graham and shot him in the back of the head with a single shot. Graham died instantaneously. Both Graham and his boyfriend are British citizens.	

'Under the Queen's Peace' and 'Malice Aforethought'

This scenario illustrates several points in murder. First, the boyfriend actually committed murder because he directly caused the death of Graham by his actions. However, Sam also committed murder as a secondary party. Second, they committed the offence of murder in Spain. Therefore, because they are British citizens they can be tried in England or Wales for the murder, no matter where it was committed in the world. This is what 'under the Queen's Peace' refers to. Finally, this case illustrates what 'malice aforethought' means. It does not mean pre-meditation. Although a lot of planning went into this particular murder by Sam and his boyfriend, murder can be an almost instantaneous action, provided the offender intends to kill when they carry out their actions. Sam and his boyfriend have the necessary specific intend to kill Graham. The intent in murder is extended a little further in that it includes an intent to cause grievous bodily harm. This means that if Sam had only meant to cause Graham grievous bodily harm, But Graham actually died from their actions, this would have been murder.

Sam also committed the offence of soliciting the murder of Graham contrary to s. 4, Offences Against the Person Act 1861:

Section 4, Offences Against the Person Act 1861

Whosoever shall solicit, encourage, persuade or endeavour to persuade, or shall propose to any person, to murder any other person, whether he be subject of Her Majesty or not ... shall be guilty of an offence.

Sam would have committed this offence even had his boyfriend not taken up the offer to kill Graham. He would also have committed the offence had he tried to engage a contract killer who in fact was an undercover police officer.

Special Defences and Voluntary Manslaughter

Murder carries a mandatory life sentence and the judge has no discretion but to sentence those convicted of murder to life imprisonment. To allow for mitigating circumstances the Homicide Act 1957 creates three unique defences that only apply to murder;

- diminished responsibility (s. 2)
- provocation (s. 3)
- suicide pact (s. 4)

In these cases the defendant admits they intended to kill, but did so only because of one of the circumstances within the defence. Any of these defences could be put forward by the accused at court. It would then be for the accused to prove that defence on the 'balance of probabilities'. Where they successfully prove they suffered from diminished responsibility, were provoked or were part of a suicide pact, they would be entitled to be found guilty of voluntary manslaughter. Voluntary manslaughter carries a maximum of life imprisonment but the judge can decide the actual sentence passed, which is often very much less than life because of the mitigating circumstances of the defence.

Involuntary Manslaughter

Involuntary manslaughter is the killing of another person but without the specific intent to kill or cause grievous bodily harm.

Common Law

Manslaughter is the unlawful killing of another human being.

Involuntary manslaughter can be committed in two ways; unlawful act (constructive manslaughter) or gross neglect.

When a person:

- kills another by an **unlawful act** which was likely to cause bodily harm:
  - there must be an unlawful act
  - the act must involve a risk of somebody being harmed
  - the defendant must have the required 'mens rea' for the relevant unlawful act.

or

- kills another by **gross negligence**:
  - … whether, having regard to the risk of death, the conduct of the defendant was so bad in all the circumstances as to amount to a criminal act or omission

Now work through the next series of case scenarios. Decide whether a manslaughter offence has been committed. If you decide an offence has been committed, consider whether a court finding of voluntary manslaughter is appropriate or whether the offence of involuntary manslaughter has been committed. Again, give reasons for your decision. The answer is provided immediately after each activity.

Activity

Case scenario	Decision and reasons
Claire and her younger brother were playing with their father's shotgun. Claire as part of the game pointed the weapon at her brother and shouted 'bang'. Her brother thought it was great fun, fell over and pretended to be dead. Claire went over to her brother while he was on the floor and again pointed the weapon at him and pulled the trigger. Unbeknown to Claire the shotgun was loaded and her brother died from the resulting injuries.	

Claire does not commit the offence of involuntary manslaughter. The unlawful act in this case was the common assault of pointing the weapon at the brother. However, Claire did not have the necessary intent to commit the assault and what resulted was a tragic accident.

Activity

Case scenario	Decision and reasons
Anne was late for an appointment and was breaking the speed limit by driving at 50 mph in her sports car. She lost control of the car as she went around a sharp bend and the car mounted the pavement. Her car hit an elderly pedestrian and killed him outright.	

Anne does not commit the offence of involuntary manslaughter. The act of driving is not in itself unlawful. The way that someone drives cannot make it an unlawful act. For this reason there are specific offences of causing death by poor driving standards. However, on rare occasions there may be exceptions to this, for instance where a person commits an assault with their car by deliberately driving their car at the person.

Activity

Case scenario	Decision and reasons
Justin was truanting from secondary school. He was messing about with some mates on a footbridge that crossed a mainline railway line. As an act of bravado Justin picked up a large boulder and threw it over the bridge just as a train was passing under the bridge. The boulder crashed through the driver's cab windscreen and struck the train driver full in the face. The driver died from the resulting injuries.	

Justin committed the offence of involuntary manslaughter by carrying out an unlawful act. He carried out the unlawful act of criminal damage. The act involved a risk of somebody being hurt as judged by a reasonable person. He had the required mens rea to commit criminal damage.

Activity

Case scenario	Decision and reasons
Dr Holloway was acting as the anaesthetist at the local hospital in a straightforward operation on a patient. He put the patient under a general anaesthetic and, contrary to standard procedures, left the operating theatre to take a smoke break. Whilst he was away, the tube supplying oxygen came loose and as a consequence the patient died. The judge at his trial decided that his actions amounted to gross negligence. The jury decided that in all the circumstances Dr Holloway's conduct amounted to a criminal omission.	

Dr Holloway commits the offence of involuntary manslaughter by gross negligence. The judge determined his conduct amounted to gross negligence and the jury decided his conduct was so bad it amounted to a criminal omission.

Activity

Case scenario	Decision and reasons
Neil and Gina were unable to continue their relationship because both their families opposed it on religious grounds. Consequently, they agreed to commit suicide together. Neil obtained a sharp-bladed knife and cut Gina's wrist for her and then hanged himself from a ceiling joist. Shortly afterwards, Neil's father discovered the two and resuscitated his son but found Gina dead.	

Neil could be found guilty of voluntary manslaughter. Neil intended to kill Gina when he cut her wrist, but then intended to kill himself. He would be guilty of the murder of Gina except for the special defence of being party to a suicide pact. Where he is able to prove this defence to a jury, on the balance of probabilities, he would be found guilty of voluntary manslaughter.

Neil also committed the offence of aiding another to commit suicide contrary to s. 2, Suicide Act 1961.

Section 2(1), Suicide Act 1961

A person who aids, abets, counsels or procures the suicide of another, or an attempt by another to commit suicide commits an offence.

Activity

Case scenario	Decision and reasons
Earl was a regular drugs dealer on the nightclub scene. He sold an ecstasy tablet to Keith at an all-night rave. Keith took the tablet and fell into a coma and was rushed to hospital. He never regained consciousness and died from the damage done by the ecstasy tablet when his life support system was switched off.	

Earl does not commit the offence of involuntary manslaughter. Supplying the tablet is not an 'unlawful act' for the purposes of manslaughter in these circumstances. This is because the victim themselves have broken the chain of causation, between Earl's unlawful act of supplying and the cause of death, by taking the tablet himself. Earl may have committed involuntary manslaughter had he directly assisted in the taking of the tablet, e.g. having gone with Keith and helped him get a drink in order to take the tablet.

Activity

Case scenario	Decision and reasons
Jim was the managing director of a railway company. He was directly responsible for the safety of the system through the routine maintenance of the track and rolling stock. He noticed that his bonus for the year was in danger of being reduced because profits for the company were poor due to rising maintenance costs. He arranged for the maintenance costs to be cut back knowing that this would increase the danger of a crash. He was prepared to take the risk to secure his full bonus entitlement. Shortly after he made these adjustments to the maintenance arrangements, one of the company's trains crashed due to a poorly maintained track. Two passengers were killed in the crash.	

Jim commits involuntary manslaughter by gross negligence (omission). This case illustrates what is known as corporate manslaughter and is explained below. Jim has a duty of care for the safety of his customers. He breached that duty of care by reducing the maintenance costs and increasing the risk of danger to them. The breach was so grossly negligent that Jim could be deemed to have had such disregard for the life of the passengers as to deserve criminal punishment.

Corporate Manslaughter

Companies have their own legal identity and can commit criminal offences. However, where the offence needs the proof of the guilty mind it is very difficult to prove against any one individual in the company who is 'directing operations'.

However, an individual may be convicted of involuntary manslaughter through gross neglect, without proving their particular state of mind, if it can be shown that:

- he defendant owed the victims a duty of care;
- the defendant breached that duty of care; and
- the breach was so grossly negligent that the defendant could be deemed to have had such a disregard for the life of the deceased as to deserve criminal punishment.

Summary

At this stage you are unlikely to be personally responsible for the investigation of murder but you may be given the task of interviewing a murder suspect. This overview may help you in the planning and preparation for the interview.

However, you may get more general responsibility in a manslaughter investigation. The legislation surrounding involuntary manslaughter can be involved and will often require the advice of the CPS as the investigation develops.

Self-assessment Test

1. Because the offence of murder has a mandatory life sentence, there are three special defences that apply only to this offence. Where the defendant proves one of these defences they are entitled to be convicted of voluntary manslaughter.

   With regard to the Homicide Act 1957, which of the following is **not** a special defence to the offence of murder?
   (a) Suicide pact.
   (b) Duress.
   (c) Diminished responsibility.
   (d) Provocation.

2. Raoul was a leader of a street gang. To prove his standing in the gang, he decided that he would set an example. He arranged a meeting with a rival gang leader in a prominent public place. When the time was right and for maximum impact he pulled out a knife and stabbed the rival gang leader in the leg intending to cause him really serious injury. The knife sliced through the main artery of the leg and the rival gang leader bled to death.

   With regard to the offence of murder, which of the following statements is correct?
   (a) Raoul did not commit murder because there was no 'malice aforethought'.
   (b) Raoul did not commit murder because he did not intend to kill the rival gang leader.
   (c) Raoul committed murder because grievous bodily harm means really serious injury.
   (d) Raoul committed murder because it was premeditated and planned.

3. Jack flew to Miami because he had been ripped off by his USA cocaine supplier over a consignment of drugs. He bought a shotgun legally from a retailer in the city centre. He then waited for the right moment and shot his supplier at point-blank range in the back intending to kill him. Jack immediately returned to this country to avoid detection. The supplier was taken to hospital and survived on a life support machine for 18 months before he died of his injuries sustained in the shooting.

   With regard to the offence of murder, which of the following statements is correct?
   (a) Jack can be tried in this country for the murder in the USA because he is a British citizen.
   (b) Jack can not be convicted of murder because the supplier died more than one year and a day after the shooting.
   (c) Jack can only be tried in the USA because the murder victim was a US citizen.
   (d) Because the victim had died more than a year and a day after the shooting, the permission of the Attorney-General would be needed to bring a prosecution.

4. Beth was heavily pregnant with Joe's baby. There was a long history of domestic violence in their marriage. Joe would often come home drunk and assault Beth. On the last occasion he was drunk he arrived home in a rage. He took a knife from the kitchen and stabbed Beth in the abdomen intending to cause her really serious injury. He had in fact forgotten that she was pregnant. As a result of the injury the baby Beth was carrying died in the womb and had to be delivered by way of an operation. Beth made a full recovery.

   With regard to the offence of murder, which of the following statements is correct?
   (a) Joe murdered the unborn baby even though he did not mean to kill Beth.
   (b) Joe did not commit murder because, in law, the unborn baby was not a human being.
   (c) Joe murdered the unborn baby because he intended to cause grievous bodily harm to Beth. Therefore, by transfer malice he intended to cause the unborn baby grievous bodily harm.
   (d) Joe committed murder because the unborn baby was, in law, a human being after 24 weeks of the pregnancy.

5. Involuntary manslaughter and voluntary manslaughter are two types of homicide. With regards to manslaughter, which of the following statements is correct?
   - (a) Voluntary manslaughter is committed when a person kills another person by gross negligence.
   - (b) Involuntary manslaughter is a finding of a court based upon any of the three special defences.
   - (c) A finding of voluntary manslaughter can result from a defence of provocation in a case of murder.
   - (d) Involuntary manslaughter is when the defendant kills another by an unlawful act even though the act does not risk anybody being harmed.

Answers to Self-assessment Test

1. (b) The three special defences under the Homicide Act 1957 can only be pleaded in a case of murder. This does not stop a defendant putting forward other defences, for example self-defence or accident. The only defence that cannot be pleaded in a case of murder is the defence of duress.

2. (c) The intention needed for murder is an intention to kill or an intention to cause grievous bodily harm. 'Malice aforethought' is in fact an old-fashioned term that does not really apply today. The term actually implies a need for premeditation. There is no requirement for premeditation in murder, the intent to kill can be almost instantaneous before the assault.

3. (a) A British citizen can be tried in this country for a murder committed anywhere elsewhere in the world. It does not matter that the victim is not a British citizen. The murder can in fact be tried in either country. Normally, other general offences committed abroad can not be tried in this country. The year and a day rule no longer exists but where a person dies three years after the injury, then the consent of the Attorney-General is required to proceed.

4. (b) Murder is the killing of another human being. In law, a baby is not a human being unless it has been born live and had an independent existence from it's mother. Transfer malice does not apply to unborn children.

5. (c) Voluntary manslaughter arises in a murder case where the accused admits killing a person but puts forward, on the balance of probabilities, one of the special defences of diminished responsibilities, provocation or suicide pact. Involuntary manslaughter is where it is accepted that the accused did not intend to kill but did kill another person because of an unlawful act or gross negligence. For involuntary manslaughter, the act must involve a risk of somebody being harmed.

Public order offences

Objectives

With regard to the most serious public order offences (Public Order Act 1986), at the end of this section, from a given set of circumstances, you will be able to:

1. Identify the points to prove for riot (s. 1(1), Public Order Act 1986).
2. Identify the points to prove for violent disorder (s. 2(1), Public Order Act 1986).
3. Identify the points to prove for affray (s. 3(1), Public Order Act 1986).
4. Identify the points to prove for intentionally causing harassment, alarm or distress (s. 4A, Public Order Act 1986).
5. Identify the offences of harassment (ss. 1, 2 and 4, Protection from Harassment Act 1997).
6. Identify the powers of injunction and restraining orders (Protection from Harassment Act 1997).

Introduction

Where there has been any large scale public disturbance, be it a 'riot' or pub fight, you will often be involved in subsequent enquiries. This can take the form of a major crime enquiry, e.g. the Broadwater Farm riot in London during the mid-1980s saw only 13 people arrested on the night whilst follow up enquiries led to a further 240 arrests. Where adequate evidence exists to charge an individual with specific offences, e.g. murder, wounding, etc., it is normal practice to charge that offence.

However, in confused situations it is difficult to link all the participants to a particular offence; there may be evidence to show some involvement, e.g. photographs showing missile throwing, but no evidence as to what the result was.

This section deals with riot, violent disorder and affray with a view to alternative charges when substantial evidence of a particular offence has not been forthcoming, as well as offences causing harassment.

Case Study 5

News of the assault on the Asian youths outside the pub quickly spread in the immediate vicinity of the town centre. It became known that the hard core of political extremists were still in the pub and were bragging about sorting out the racial problems in the area. Two local Asian men approached the pub and stood on the pavement opposite. One started shouting into the pub, 'You are scum and we don't want you here. You are not going to get your way. Go and go now before you regret it.' The two then continually shouted 'Scum! Scum! Scum!'

towards the pub. This went on for a few minutes during which one or two of the extremists made some defiant obscene gestures at the men through the pub windows.

The two local men realised that the extremists clearly intended to stay put. The local men became more animated and shaped up as if they wanted to fight. They shouted 'Alright, if you want trouble you can have it. If you bastards want it come out here and you can have some. Let's sort it. If you don't come out, we're coming in.'

A small crowd of six or seven Asian men joined the two local men. All the men now joined in the shouting and started towards the pub entrance clearly intent on violence. On seeing the crowd outside the pub on the move, the extremists decided to make a run from the pub. Some of the crowd caught one of the fleeing extremists and started to kick him while he was on the pavement.

More and more Asian men joined in the disorder until a large crowd of over 50 people was attacking the pub. About six of the men armed themselves with iron bars and house bricks and started to break windows and damage the building wherever they could. The rest of the crowd was with them cheering and gesturing. Eventually one of the group threw a petrol bomb through a window and started a fierce fire inside.

From the circumstances described, identify what offences have been committed contrary to public order legislation.

....

....

....

....

....

....

You should have noted that, to start, the two Asian men commit an offence of intentionally causing harassment, alarm and distress (probably in the racially aggravated form). They then go on to commit an affray. Once three or more people are involved, violent disorder is committed. Finally, provided a common purpose is proven, a riot would have been committed.

Look now at each offence in turn, starting with the most serious.

Riot

Section 1(1), Public Order Act 1986

Where 12 or more persons who are present together use or threaten unlawful violence for a common purpose and the conduct of them (taken together) is such as would cause a person of reasonable firmness present at the scene to fear for his personal safety, each of the persons using unlawful violence for the common purpose is guilty of riot.

Only the person or persons who actually use violence in the above circumstances commit the offence of riot. It is not necessary that all members of the group use violence, but the others must be 'threatening'. These 'others' do not commit riot but could be guilty of a lesser offence, e.g. violent disorder. Without someone actually using violence there can he no 'riot'.

Violence includes conduct intending to cause injury to the person or damage to property and other violent conduct, e.g. missile throwing which does not hit or falls short. The violence must be 'unlawful', so 12 or more police officers acting together baton charging a crowd in the execution of their lawful duty to restore the Queen's peace do not commit the offence of riot.

'Present together' means gathered at the same location. It does not mean that the participants all have to be part of the same group, e.g. several gangs of five or six people all gathered in a market square would be '12 or more present together'.

The 'common purpose' may be unlawful, e.g. to attack the police or the shop premises of local traders. It could also be lawful, e.g. to lobby members of the local council. Individual motives for joining in may have nothing to do with this 'common purpose'. This is immaterial as it will have to be proved from inferences based on the general conduct of the 12 or more.

No person of reasonable firmness need actually be, or be likely to be, present at the scene. All that is required is that the conduct of the 'rioters' would cause such a hypothetical person to fear for their personal safety, presumably the fear would have to be for his immediate personal safety. Riot may be committed on private premises as well as public places.

Riot is triable on indictment only with a maximum of 10 years' imprisonment. Prosecution for riot requires the consent of the Director of Public Prosecutions.

Violent Disorder

Section 2(1), Public Order Act 1986

Where 3 or more persons who are present together use or threaten unlawful violence and the conduct of them (taken together) is such as would cause a person of reasonable firmness present at the scene to fear for his personal safety, each of the persons using or threatening unlawful violence is guilty of violent disorder.

The offence has many similarities to riot. However, there are differences, notably that there need only be '3 or more persons' and that there is no requirement that an accused actually use violence to be guilty of the offence. Threatening violence will be sufficient for violent disorder.

No reference is made to 'common purpose' so it may be that large scale unrest will be prosecuted as violent disorder, not riot, except in the most serious cases or in the clearest cases of common purpose.

As with riot, violent disorder can take place on private premises, as well as in a public place.

Violent disorder is triable either way with a maximum punishment of five years' imprisonment on indictment.

Affray

Section 3(1), Public Order Act 1986

A person is guilty of affray if he uses or threatens unlawful violence towards another and his conduct is such as would cause a person of reasonable firmness present at the scene to fear for his personal safety.

This offence can be committed by an individual acting alone. The violence here cannot be directed against property but must be directed against a particular person or persons. In circumstances where violence is threatened, the threats must be accompanied by threatening gestures. Where two or more persons use or threaten unlawful violence, it is the conduct of them taken together that must be considered.

Like riot and violent disorder the violence must be unlawful, so a person fighting in self defence cannot be guilty of affray. The hypothetical person of reasonable firmness need not be present, however this requirement would appear to exclude some fights from the definition of the offence, e.g. a fight between two people arising out of a personal quarrel and unlikely to involve other people would not be affray as it would not cause another to fear for his personal safety. Affray can be committed on private premises as well as in public.

Affray is triable either way with a maximum punishment of three years' imprisonment. There is a power of arrest for a constable where he or she reasonably suspects anyone of committing affray.

Intentionally Causing Harassment, Alarm or Distress

You will be aware of offences under s. 5 of the Public Order Act 1986. An offence has been created which addresses loopholes in s. 5 and also widens police powers.

Section 4A, Public Order Act 1986

A person is guilty of an offence if, with intent to cause a person harassment, alarm or distress, he—

(a) uses threatening, abusive or insulting words or behaviour, or disorderly behaviour, or

(b) displays any writing, sign or other visible representation which is threatening, abusive or insulting,

thereby causing that or another person harassment, alarm or distress.

The only difference between this offence and s. 5 of the Public Order Act 1986 is that there has to be an *intention* to cause a person harassment, alarm or distress.

Harassment, alarm or distress are not defined and although the section was brought in to address the problem of racial harassment, it also covers other types of harassment including sexual and religious.

An offence can be committed in public and private but not in dwellings (to exclude domestic disputes); for example, pickets threatening work colleagues whether inside or outside the factory, or protesters who invade a military base provided all the elements are satisfied.

A police officer may arrest without warrant anyone whom he or she reasonably suspects *is committing* an offence under s. 4A. A police officer does not have to be in uniform to exercise this power.

As usual, where the offence has been committed, the general power of arrest under s. 25, Police and Criminal Evidence Act 1984 could be used.

The offence is triable summarily only, with a maximum punishment of six months' imprisonment and/or a fine. (No imprisonment with s. 5.) The penalty is increased to two years' imprisonment if racially or religiously aggravated (s. 31(1)(b), Crime and Disorder Act 1998).

The Mental Element

The *mens rea* for riot, violent disorder and affray requires an intention by the accused to use or threaten violence or that he or she is aware that his or her conduct may be violent. 'Aware' is a concept similar to recklessness. Even if a person is intoxicated, he or she is still regarded as being as 'aware' as if they were sober, the only exception being if the drink, drugs etc. were not self-induced. This is sensible given that affray in particular is a useful offence to deal with drunken people fighting outside pubs etc.

For the offence under s. 4A, similar conditions apply in relation to the intention to cause harassment, alarm or distress.

Protection from Harassment Act 1997

Offences under ss. 1 and 2, Protection from Harassment Act 1997, were introduced to deal with the problem of 'stalking' (persistent and often obsessive behaviour targeted against a particular individual).

Although targeted at stalkers, the scope of the Act is much wider and may be applied to domestic incidents, neighbourhood nuisances, bullying at work, racial/sexual harassment and even intrusive news reporters.

The provisions of the Act make it possible for police and victims to take action at an earlier stage than previously. The offences under the Act may be committed *anywhere*, they are all *arrestable* and there is no requirement to prove a specific intent. For the first time, criminal courts now have the power to control offenders' behaviour *after conviction* (of offences under the Act) by means of a restraining order.

Here are the key features of the 1997 Act:

Harassment

Sections 1 and 2 deal with basic harassment. It states that a person must not pursue a 'course of conduct' which amounts to harassment of another and which the person knows or ought to know amounts to harassment of the other.

The offence is arrestable and has the penalty of six months' maximum imprisonment that is increased to two years where the offence is racially or religiously aggravated.

Harassment includes alarming the person or causing them distress. For a 'course of conduct' you would need to prove that the behaviour was conducted on at least two occasions. Behaviour like sending flowers to a person could amount to harassment.

Putting People in Fear of Violence

Section 4 provides the more serious offence of putting people in fear of violence. Here the offence is committed when a person undertakes a 'course of conduct' that causes another to fear that violence will be used against them when the person knows or ought to know that their course of conduct would cause the other to so fear.

This offence is also arrestable and has the penalty of five years' maximum imprisonment that is increased to seven years where the offence is racially or religiously aggravated.

Injunctions

Under s. 3, a judge in a civil action in the High Court or county court may issue an injunction restricting the behaviour of a person to prevent that person harassing another. This is even before any criminal offence has been committed. The court can issue a warrant of arrest when the person fails to adhere to the restrictions described in the injunction. The person commits a criminal offence of breach of injunction that has the penalty of five years' maximum imprisonment and can be dealt with in the criminal court.

Restraining Orders

Section 5 provides a criminal court with the power to restrict the behaviour of a person who is convicted of an offence under ss. 2 and 4. This is in addition to any other penalty imposed by the court. A breach of a restraining order is also a criminal offence in its own right, having the same penalty as that for the breach of an injunction in the previous paragraph.

Summary

Public order is not normally regarded as a crime matter, however violence by large numbers of offenders will often be investigated after public order has been restored. It should not be forgotten that there is likely to be a mix of offences which can be applicable to circumstances you may be investigating.

Self-assessment Test

Having completed this section, test yourself against the objectives outlined at the beginning of the section. You will find the answers below.

1. With regard to the Public Order Act 1986, which of the following statements is *correct?*
   (a) Clive and Darren are guilty of violent disorder (s. 2(1)) when they called round a work colleague's house armed with baseball bats. They forced their way into the hallway and started shouting and threatening the work colleague, shouting 'You are going to get plenty of this' (indicating the baseball bats).
   (b) Jane had organised a pro-life rally with 11 other women outside the headquarters of a major drug company. They were all intent on bringing pressure on the company to prevent the sale of contraception pills in chemists. Jane committed riot (s. 1(1)) when she started breaking the windows of the headquarters by throwing stones and bricks. The other women were cheering Jane and shouting and screaming 'You bastards are not human, we are going to get you'.
   (c) George picked up a wooden post and smashed the windscreen of a car parked on the private driveway of a house. He shouted 'I'm fed up with this government they are only interested in wealth'. George is guilty of affray (s. 3(1)).
   (d) John was drunk in the street. He was shouting and swearing but had no intention to harass, alarm or distress anyone. John is guilty of intentional harassment, alarm or distress (s. 4A).

Answer to Self-assessment Test

1. Answer (b). Only one person needs to actually use violence, while the other 11 need to be threatening. The common purpose can be lawful.

Criminal damage

Objectives

With regards to criminal damage (Criminal Damage Act 1971), at the end of this section, from a given set of circumstances, you will be able to:

1. Identify the points to prove for criminal damage (s. 1(1), Criminal Damage Act 1971).

2. Identify the points to prove for criminal damage when life is endangered (s. 1(2), Criminal Damage Act 1971).

3. Identify 'recklessness' as outlined in *Metropolitan Police Commissioner* v *Caldwell*.

4. Identify an offence of arson (s. 1(3), Criminal Damage Act 1971).

5. Identify the points to prove for threats to destroy or damage property (s. 2, Criminal Damage Act 1971).

6. Identify the points to prove for possession of anything with intent to destroy or damage property (s. 3, Criminal Damage Act 1971).

7. Identify the mode of trial for offences under the Criminal Damage Act 1971.

8. Identify an offence of contamination or interference with goods (s. 38, Public Order Act 1986).

Introduction

The introduction of the Criminal Damage Act 1971 repealed the Malicious Damage Act 1861 except for six sections which relate to offences against rail transport and shipping buoys. We will not be discussing either in this section. In this section we will be looking at various aspects of criminal damage.

Greenall and Smith have an argument in a pub, Greenall was so outraged that she went to Smith's home and poured brake fluid over his motor car. From the facts given, what offence(s) has Greenall committed in relation to the damage to the vehicle?

....

....

....

....

....

....

Your answer should be criminal damage contrary to s. 1(1), Criminal Damage Act 1971. You may also have identified possession of an article to commit criminal damage contrary to s. 3, Criminal Damage Act 1971. We will discuss s. 3 later, but for now let us look at s. 1(1).

Section 1(1), Criminal Damage Act 1971

A person who without lawful excuse destroys or damages any property belonging to another intending to destroy or damage any such property or being reckless as to whether any such property would be destroyed or damaged shall be guilty of an offence.

There are several terms in that definition we need to explore, in particular, *destroy* and *damage*.

(a) *Destroy*—clearly goes beyond damage, indicating a damage that is beyond repair.

(b) *Damage*—there is no need for property to be permanently damaged, only that what is done causes some 'temporary functional derangement', e.g. removing a part from a machine rendering it inoperative is damage even though it can readily be repaired by replacing the missing part.

A car is equally damaged by uncoupling a brake cable as it is by severing it with a saw or a pair of pliers.

It has been held that spitting on a police officers's raincoat was not damage as the spittle could be removed with the wipe of a cloth. However, in another case, a police officers's cap was jumped upon and this was found to be criminal damage despite there being no evidence that the cap might not have been restored to its original state without any real cost or trouble to the owner.

It is therefore difficult to give any precise or definitive rule as to what constitutes damage and in some cases it may be a matter of fact for a jury or magistrate to decide.

Section 5, Criminal Damage Act 1971, provides two defences to causing damage. The first defence allows for when a person believes they have or would have had permission from the owner to do the damage. The second defence is where damage is done in the honest belief that it was necessary to protect other property.

The criminal damage to the pub windows in our public order case study appeared to be racially motivated. In this case, where the offence is proven at court to be racially or religiously aggravated, the penalty increases from ten to fourteen years' imprisonment.

Damage to Own Property

Let us imagine that Greenall and Smith in fact know each other well and that Greenall cut the brake pipe of her own car and then lent the vehicle to Smith knowing he was setting off on a long journey.

What offence under the Act has Greenall committed in these circumstances? Write your answer quoting the section below.

....

....

....

....

....

....

You will have realised that no offence is committed under s. 1(1) as the property is damaged by the owner of it. This time the offence is covered by s. 1(2) of the Act.

Look at the full definition below.

Section 1(2), Criminal Damage Act 1971

A person who without lawful excuse destroys or damages any property, whether belonging to himself or another—

(a) intending to destroy or damage any property or being reckless as to whether any property would be destroyed or damaged; and

(b) intending by the destruction or damage to endanger the life of another or being reckless as to whether the life of another would be thereby endangered;

shall be guilty of an offence.

An important aspect of this section is that to damage or destroy your own property is an offence, if life would be endangered.

Recklessness

Recklessness is when a person takes an unjustified risk. The concept has been confusing in the past because criminal damage alone applied an objective test for recklessness (what would the reasonable man have thought and done). Recklessness for criminal damage is now a subjective test as a result of a recent House of Lords case (*R* v *G&R* (2003)), which sets out the test for recklessness.

Their Lordships held that a person acts recklessly for the purposes of s. 1(1) Criminal Damage Act 1971;

- with respect to the circumstances when s/he is aware of a risk that existed or would exist;

- with respect to a result or consequence when s/he is aware of a risk that it would occur and it is, in the circumstances known to him/her, unreasonable to take the risk.

In this particular case two children were convicted of starting a fire which caused around £1m of damage. The conviction was based upon their recklessness judged against the previous objective test. The House of Lords quashed the conviction and replaced the objective test.

Damage by Fire

If an offence contrary to s. 1(1) or (2) is committed, and the means of the destruction or damage is by fire, the offence is contrary to the relevant section and contrary to s. 1(3).

Section 1(3), Criminal Damage Act 1971

An offence committed under this section by destroying or damaging property by fire shall be charged as arson.

You may be wondering what this means. Well consider the first set of circumstances where Greenall had poured brake fluid on the car. If instead she had set fire to it, the offence would have been criminal damage contrary to s. 1(1) and 1(3), Criminal Damage Act 1971.

In the second set of circumstances, where Greenall had cut the brake pipe of her own car to endanger Smith's life, had she set fire to her own car while Smith was in it, the offence would have been criminal damage contrary to s. 1(2) and 1(3), Criminal Damage Act 1971.

Threats to Destroy or Damage Property

Let us stick to the first two sets of circumstances involving the damage caused by Greenall to Smith's car and her own.

Instead of causing the actual damage, what if Greenall in both cases just threatened Smith that she would cause the damage?

How many offences of 'threats to damage' would have been committed?

....

....

....

....

....

....

The answer is that an offence of 'threat to destroy or damage property' contrary to s. 2(a) or (b), Criminal Damage Act 1971 has been committed in both cases.

Look at the full offence.

Section 2, Criminal Damage Act 1971

A person who without lawful excuse makes to another a threat, intending that that other would fear it would be carried out—

(a) to destroy or damage any property belonging to that other or a third person; or

(b) to destroy or damage his own property in a way which he knows is likely to endanger the life of that other or a third person;

shall be guilty of an offence.

Therefore, you can commit an offence of 'threats to damage property' even though you are threatening to damage your own property. Again 'likely to endanger the life' is an important element in that case.

The victim of the threat does not have to believe the threat would be carried out; it is for the person threatening to *intend* that the victim would fear it to be carried out.

Possession of Anything to Cause Damage

See how much you can remember from your previous training by doing the next activity.

Answer the following questions which relate to possession of items to cause criminal damage.

1. What type of items are included in the Act to make it an offence to possess items to cause criminal damage?

....

....

....

....

2. Where must the items be for the offence to be committed?

....

....

....

....

3. What sort of damage must be intended with such possession?

....

....

....

....

You can find the answers in the definition of the offence.

Section 3, Criminal Damage Act 1971

A person who has anything in his custody or under his control intending without lawful excuse to use it or cause or permit another to use it—

(a) to destroy or damage any property belonging to some other person; or

(b) to destroy or damage his own property or the user's property in a way which he knows is likely to endanger the life of some other person;

shall be guilty of an offence.

You should see that the answer to the first question is 'anything'. The answer to the second question is 'anywhere', provided it is in the person's custody or control. The answer to the third question is possession with intention to cause 'any damage by the person or another referred to in s. 1 of the Act'.

Have a look now at two definitions you need to know:

'Property' and 'Belonging to Another'

'Property' is basically the same as for theft, except in two areas:

1. You can damage land, whereas generally you cannot steal it.

2. You cannot damage intangible property or things in action, whereas you can steal them.

As you can see these are logical rules. You can damage your neighbour's lawn in many ways, but cannot steal it by moving your garden fence over a bit. You can steal the space in your neighbour's builder's skip, but cannot damage that space.

'Belonging to another' in criminal damage is similar to belonging to another in theft. This is having the custody or control of the property, or a proprietary right or interest in it. But in criminal damage it goes a bit further; 'belonging to another' also includes anyone with a *charge* on the property.

So the victim of damage does not actually have to own the property. For instance, Barclays Bank could be a victim if they have a charge over a damaged house even though the Bank is not the owner of it.

Thus 'property' includes land but excludes intangible property; 'belonging to another' includes a person who has a charge on the property.

Mode of Trial

There are many types of damage that can be caused. This ranges from the most trivial to the very serious. So the place of trial for a particular offence will depend on the seriousness of the damage and the intent behind it.

Have a look at the circumstances in this table and test your knowledge by identifying where each would be tried. Tick the appropriate column.

Circumstances	Where offences could be tried		
	Crown Court only	Magistrates' court only	Either Crown or magistrates' court
£4,000 worth of criminal damage to a motor car by the spraying of aerosol paint.			
£50 worth of criminal damage to a motor car by fire,			
£15 worth of criminal damage to a gas pipe with intent to endanger life.			
£5,500 worth of criminal damage to the glass of a shop window.			
£2,200 worth of criminal damage to a garden fence by ripping it down.			
£700 worth of criminal damage each in a series of eight offences to shop windows.			

Look at the following breakdown of offences and modes of trial:

1. Crown Court only. Any criminal damage with intent to endanger life.
2. Magistrates' court only. Any criminal damage except by fire, where the value of the damage does not exceed £5,000.
3. Either at Crown Court or magistrates' court. All criminal damage where the value is in excess of £5,000 or under £5,000 if in a series of such offences where the aggregate value of the series exceeds £5,000 and all criminal damage by fire irrespective of the value.

The offences of 'threats to commit damage' and 'possession of articles to commit damage' are triable in either court regardless of the intent.

Contamination or Interference with Goods

Section 38, Public Order Act 1986, creates three offences. The first offence is the contamination of goods with intent to cause public alarm, injury or economic loss (s. 38(1)). The second offence is the threat to contaminate goods as above (s. 38(2)). The third offence is to have possession of articles that are contaminated or articles that are used to contaminate goods (s. 38(3)).

These offences are often accompanied by blackmail. The offences carry a penalty of 10 years' imprisonment and are arrestable.

Summary

In this section we have dealt very basically with offences under the Criminal Damage Act 1971. As with assaults and public order, criminal damage is a common type of offence that is committed against ethnic minority groups. In the multi-racial and multi-cultural society that we police in England and Wales, as investigators we must be aware of and understand the elements of racially or religiously motivated incidents.

Self-assessment Test

Having completed this section, test yourself against the objectives outlined at the beginning of the section. You will find the answers below.

1. With regard to s. 1(1), Criminal Damage Act 1971 which of the following statements is *incorrect*?
   (a) Section 5, Criminal Damage Act 1971 provides a defence to causing damage where the person causing the damage honestly believed that it was necessary to cause the damage to protect other property.
   (b) The inclusion of the word 'destroy' in the wording of the offence means that any damage to property must be permanent and beyond repair for an offence to be committed.
   (c) Where it is proved at court that the criminal damage is racially motivated, the maximum penalty increases from 10 to 14 years' imprisonment.
   (d) A defendant is reckless with respect to the circumstances when s/he is aware of a risk that existed or would exist and with respect to a result or consequence when s/he is aware of a risk that it would occur and it is, in the circumstances known to him/her, unreasonable to take the risk.

2. With regard to s. 1(2) and (3), Criminal Damage Act 1971, which of the following is *correct*?
   (a) The offence of criminal damage with intent to endanger life will always involve damage caused by fire.
   (b) Arson is an offence where the damage caused can only be damage to property belonging to another person.
   (c) A person can be reckless as to whether the life of another would be endangered.
   (d) Criminal damage with intent to endanger life is an offence where the damage caused can be damage only to property belonging to another person.

3. Keith found out that his partner was regularly seeing Mark. He became jealous and decided to frighten Mark to break up the relationship. Keith bought some petrol in a petrol can and kept it in the garage at his house, intending to start a fire at Mark's flat. While Keith was at work he telephoned Mark and said 'Unless you keep yourself to yourself I will torch your flat'. Keith knew the fire was likely to endanger life.

   With regard to ss. 2 and 3, Criminal Damage Act 1971, which of the following is *correct*?
   (a) Keith commits the offence of possession of the petrol intending to damage Mark's property because he knew the fire was likely to endanger life.
   (b) Keith does not commit the offence of possession of the petrol to damage Mark's property because he did not have the petrol with him when he made the threat.

(c) Keith commits the offence of threatening to damage property belonging to another because Keith intended to cause the damage.

(d) Keith does not commit the offence of threatening to damage property belonging to another because Mark did not believe that the threat would be carried out.

Answers to Self-assessment Test

1. Answer (b). Property can be either destroyed (beyond repair) or damaged (but repairable).

2. Answer (c). A person can intend to commit damage or be reckless as to whether damage is caused and intend to endanger life or be reckless as to whether life is endangered.

3. Answer (a). Keith must know his actions are likely to endanger life.

Misuse of drugs

Objectives

With regard to the misuse of drugs, at the end of this section, from a given set of circumstances, you will be able to:

1. Identify to which class of drug the most common controlled drugs belong (Schedule 2).
2. Identify where the offences of importation and exportation are found (Customs and Excise Management Act 1979).
3. Identify the points to prove for the production of a controlled drug (s. 4(2)).
4. Identify the points to prove for the cultivation of cannabis (s. 6).
5. Identify the points to prove for the supplying of a controlled drug (s. 4(3)).
6. Identify the points to prove for possession with intent to supply a controlled drug (s. 5(3)).
7. Identify the points to prove for supplying articles for administering or preparing controlled drugs (s. 9A).
8. Identify the points to prove for the occupier or manager of premises permitting drug misuse (s. 8).
9. Identify the points to prove for the possession of a controlled drug (s. 5(2)).
10. Determine when the defence to the offence of possession applies (s. 5).
11. Identify who is exempt from prosecution for offences against the Misuse of Drugs Act 1971 (Misuse of Drugs Regulations 2001).
12. Determine when the general defence to other offences (s. 28).
13. Identify the points to prove for incitement (s. 19).
14. dentify the points to prove for assisting or inducing misuse of drugs offences abroad (s. 20).
15. Identify the powers of stop, search, detain, entry, search and seizure (s. 23).
16. Identify the points to prove for obstruction (s. 23(4)).
17. Distinguish between the penalties for offences under the Misuse of Drugs Act 1971.
18. Recognise when a urine sample or non-intimate sample can be taken from a detained person to test for Class A drugs (Criminal Justice and Court Services Act 2000).

Introduction

In this section we will be restricting ourselves to look at the law relating to drugs and in particular those aspects which may involve you.

Since the 1960s the misuse of drugs in the western world has been an increasing problem which shows no sign of abating. The Misuse of Drugs Act 1971 set out to consolidate the piecemeal legislation which had grown up since the turn of the century with a view to controlling the use of drugs in every connection and provide the facility to make regulations to deal with future developments in an ongoing situation.

Classification of Controlled Drugs

The 1971 Act does not give a definition of a 'drug' or 'narcotic' but it says that a 'controlled drug' means any substance or product for the time being specified in Schedule 2 of the Act.

In Schedule 2, controlled drugs are divided into three classes according to their toxic effect, prevalence of misuse and the danger to society. The only reason they are divided is for determining which penalties may be imposed for offences involving their misuse.

The list of controlled drugs is lengthy and for practical purposes it is not necessary to be aware of every one as, when making an arrest, the criteria will be 'reasonable suspicion' that the substance is a controlled drug.

They are classified as follows:

Class A	Includes all the well-known addictive drugs eg morphine, opium, heroin, cocaine, injectable amphetamines, LSD and cannabinol (except where contained in cannabis or cannabis resin).
Class B	Includes some amphetamines and codeine.
Class C	Includes milder stimulants, cannabis and cannabis resin.

Those are the classifications, now we will look at the offences.

Case Study 6

Mack had been on the fringes of the drugs world for several years. He always kept a small amount of cocaine and cannabis in his bedroom for his use at weekends. Just recently he had realised that his cocaine use had developed into a serious habit and he was finding he needed more and more to get the same effect. He was struggling to pay for the drugs he now needed. Mack noticed that there was a gap in the local drugs market. He decided to approach his usual supplier and arrange to buy enough cocaine so that he could set himself up as a small dealer in his local pubs. He phoned his supplier Oscar and said, 'When can you get me the stuff I told you I needed?'. Oscar replied, 'Its okay, its coming in soon but there's a shortage due to improved security at the ports. I'll ring you when it's here'. Oscar took delivery of a large assignment of cocaine from Spain. Mack and Oscar met at a motorway services and Oscar gave Mack a large quantity of cocaine in exchange for money he had borrowed. Mack went home and cut the drug with some baking powder, divided the mixture up into saleable wraps and went to work. Mack set himself up in the White Horse pub to start with because he knew the pub manager was lax and the demand would be high. The manager of the pub knew Mack used drugs and now that Mack was hanging around all the time, without drinking very much he was pretty sure he was dealing in the toilets of the pub. He ignored Mack's dealing because he knew Mack could be violent and didn't want any trouble from him. Mack sold his cocaine from the toilets of the pub. Mack asked a friend, Katie, to look after some of his cocaine for a couple of days as a favour. She gave it back to Mack when he asked for it. After a short while, business was so good that Katie agreed to help him deal in the pub because she needed drugs herself. Katie went up to a female customer in the pub and said, 'Do you want some Charlie?'. The female customer asked about the quality and price and agreed to buy some. Katie sent her to the toilets and in the lobby area Mack sold the customer a wrap of cocaine.

From these circumstances and ignoring conspiracy, list the possible offences committed contrary to drugs legislation, by each person mentioned.

Person mentioned	Offence committed	Act and section
Mack		
Oscar		
Pub manager		
Katie		
Female customer		

Your table should have looked like this:

Person mentioned	Offence committed	Act and section
Mack	Possession in his bedroom. Inciting Oscar to supply. Inducing offence abroad. Possession with intent to supply cocaine. Supplying cocaine. Inciting Katie to be concerned.	s. 5(2),Misuse Drugs Act 1971 s. 19 s. 20 s. 5(3) s. 4(3) s.19
Oscar	Assisting offence abroad. Importation. Possession with intent to supply cocaine. Supplying cocaine.	s. 20 s. 3 and Customs and Excise Management Act 1979. s. 5(3) s. 4(3)
Pub manager	Permits supplying.	s. 8
Katie	Possession of cocaine. Offering to supply cocaine. Concerned in supply cocaine.	s. 5(2) s. 4(3) s. 4(3)
Female customer	Possession of cocaine.	s. 5(2)

You may have had slight variations but one thing you will notice is that the legislation is very wide-ranging. It covers virtually any unlawful transaction where a controlled drug is concerned. Let's look at offences of production, supplying and possession in turn and bring in other sections as they become relevant.

Unlawful Importation and Exportation

Section 3, Misuse of Drugs Act 1971

Prohibits the importation and exportation of controlled drugs in and out of the UK unless authorised by regulations under the Act.

Section 3 does not create anzy offences. Unlawful importation and exportation of goods are offences under the Customs and Excise Management Act 1979. All s. 3 does is to include controlled drugs as goods subject to the control of that Act.

Unlawful Production

Section 4(2), Misuse of Drugs Act 1971

It is an offence for a person to unlawfully:

(a) produce a controlled drug

(b) be concerned in the production of a controlled drug.

Produce is defined by Section 37 as meaning producing it by manufacture, cultivation or any other method and 'production' has a corresponding meaning.

Converting one form of a Class A drug into another is producing. Production would include harvesting, cutting and stripping a cannabis plant.

Cultivation of Cannabis

Section 6(2), Misuse of Drugs Act 1971

It is an offence unlawfully to cultivate any plant of the genus Cannabis.

Cultivation requires an element of '*mens rea*' and, to prove an offence, some degree of attention to the plants would have to be proved e.g. watering, hoeing etc.

What is Cannabis?

Section 52, Criminal Law Act 1977

Cannabis means any plant of the genus cannabis, or any part of any such plant (by whatever name called) except that it does not include cannabis resin (a Class B drug) or any of the following products after separation from the rest of the plant:

(a) mature stalk of any such plant;

(b) fibre produced from mature stalk of any such plant; and

(c) seed of any such plant.

'Cannabis resin' means the separated resin, whether crude or purified, obtained from any plant of the genus cannabis.

Unlawful Supply

Section 4(3), Misuse of Drugs Act 1971

It is an offence for a person unlawfully:

(a) to supply a controlled drug to another.

(b) to be concerned in the supplying of such a drug to another.

(c) to offer to supply a controlled drug to another.

(d) to be concerned in the making to another of an offer to supply such a drug.

'Supplying' includes distributing (s. 37), therefore, to buy controlled drugs for oneself and others is supplying, because distributing is supplying.

In our case study, Oscar clearly supplies cocaine to Mack, who then later supplies it to the female customer in the toilet area of the pub, because both Mack and the customer are able to use the cocaine in anyway they want.

However, there are several instances you need to note regarding supplying;

- There is no supply by Mack when he asks Katie to look after the cocaine for a couple of days, nor is there a supply by Katie when she gives it back to Mack.

- Had Katie benefited in some way by looking after the drugs, in that case she would commit the offence of supplying Mack on returning the cocaine.

- Had Katie been threatened and made to look after the drugs for someone she wasn't able to name and then given them back on demand she could have been guilty of supplying.

- Had Mack injected Katie with her own drugs, this would not be supplying Katie.

- Had the female customer, who Katie made the offer to, been an undercover police officer, Katie would still have committed the offence of offering to supply.

Katie commits the offence of offering to supply cocaine as soon as she makes the offer. It doesn't matter whether she can supply or intends to supply the drug. When the cocaine is supplied to the customer, Katie has been concerned in its supply.

Possession with Intent to Supply

Section 5(3), Misuse of Drugs Act 1971

It is an offence for a person to have a controlled drug in his possession.
Whether lawfully or not, with intent to supply it unlawfully to another.

This is an offence of specific intent. Having proved 'possession' of a controlled drug, what you need to prove next is an intention to supply it unlawfully to another. Usually this evidence can be adduced by a large quantity of drugs carried, of the drugs being wrapped in several small packets ready for distribution, or the person also carrying a large sum of money, or a list of customers etc.

Notice that a person who is in lawful possession of controlled drugs can commit this offence if they intend to unlawfully supply it to another.

In our case study, if it could be proved that both Mack and Katie had possession of the cocaine, you would need to prove that each of them intended to supply the cocaine to another for each to commit the offence. It would not be enough to prove that Katie simply knew of Mack's intention to supply. Interestingly, had Mack in fact had possession of heroin but thought he had possession of cocaine, he would still commit possession of the heroin with intent to supply.

You may not find it easy to prove Mack's intent if he is found with the cocaine with no other evidence. However, Mack's unexplained wealth or his possession of a large amount of cash when he is arrested with the cocaine can be given in evidence to prove his intent.

Supply of Articles

Section 9A, Misuse of Drugs Act 1971

(1) A person who supplies or offers to supply any article which may be used or adapted to be used (whether by itself or in combination with another article or other articles) in the administration by any person of a controlled drug to himself or another, believing that the article (or the article as adapted) is to be so used in circumstances where the administration is unlawful, is guilty of an offence.

(3) A person who supplies or offers to supply any article which may be used to prepare a controlled drug for administration by any person to himself or another believing that the article is to be so used in circumstances where the administration is unlawful is guilty of an offence.

There are only two offences, one under s. 9A(1), and the other under s. 9A(3), since the penalty does not vary with the controlled drug in question.

This offence deals with articles which enable people to administer controlled drugs to themselves or others. Some of these articles will be known as drug kits, but do not include hypodermic syringes or parts of them.

While in this offence what is being supplied or offered for supply is an article rather than a controlled drug, it may be that the same approach as in, for example, the Misuse of Drugs Act 1971, s. 4(3), applies to the interpretation of the phrase 'supplies or offers to supply'.

The articles must be for the unlawful administration of a controlled drug.

Section 9A, Misuse of Drugs Act 1971

(4) For the purposes of this section, any administration of a controlled drug is unlawful except—
   (a) the administration by any person of a controlled drug to another in circumstances where the administration of the drug is not unlawful under Section 4(1) of this Act, or

   (b) the administration by any person of a controlled drug to himself in circumstances where having the controlled drug in his possession is not unlawful under Section 5(1) of this Act.

Controlled Drugs on Premises

Section 8, Misuse of Drugs Act 1971

A person commits an offence if, being the occupier or being concerned in the management of any premises, he knowingly permits or suffers, any of the following activities to take place on those premises:

(a) unlawfully producing or attempting to produce a controlled drug.

(b) unlawfully supplying or attempting to supply a controlled drug to another, or offering to supply a controlled drug unlawfully to another.

(c) preparing opium for smoking

(d) smoking Cannabis, Cannabis Resin, or prepared Opium.

The manager of the White Horse pub in our case study is clearly concerned in the management of the premises Mack is using to supply drugs. When an occupier of a residential flat, business unit or other premises is committing the offence, you do not have to prove a strict legal status for the occupier. What is important is that the occupier had enough control over the premises to prevent the supplying or other activity mentioned in the offence.

The pub manager must 'knowingly' permit or allow the activity, but this includes 'wilful blindness'. This amounts to closing his eyes to the obvious or not caring what happens. In our case the manager commits the offence because he ignores the obvious. He doesn't need to know exactly what type drug is being produced, supplied etc.

Possession

This is the offence normally committed by the user.

Section 5(2), Misuse of Drugs Act 1971

It is an offence for a person unlawfully to have a controlled drug in his possession.

The things which a person has in his possession shall be taken to include anything subject to his control which is in the custody of another (s. 37).

It is possible to have more than one person in possession of the same drug, but each would need to have 'control' over the drug and also knowledge of it.

In our case study, it is for the prosecution to prove that the substance was cocaine and that Mack had possession of it because he had control of it and that he knew that a substance existed. Once the prosecution have proved this, it would be up to Mack to convince the court on the balance of probabilities that he had a defence under s. 5 (possession only) or s. 28 (possession and other offences) to have possession of the cocaine.

You would not have to show Mack knew it was cocaine he had in his possession. Neither would it matter how small a quantity Mack had in his possession provided it was 'visible, tangible and measurable'. However, where the amount is too small to prove possession of the small amount, its presence may go to prove an earlier possession of the drug.

Let's look at the defences.

There are two special defences to an offence of 'simple possession', which an accused can rely on after it has been proved that he had a controlled drug in his possession.

The Misuse of Drugs Act 1971, s. 5(4), provides a defence specifically to the possession offence in s. 5(2), but the existence of that defence does not preclude any other defences, either specific ones under the Misuse of Drugs Act 1971, or relevant general defences.

Section 5, Misuse of Drugs Act 1971

(4) In any proceedings for an offence under subsection (2) above in which it is proved that the accused had a controlled drug in his possession, it shall be a defence for him to prove:-

(a) that, knowing or suspecting it to be a controlled drug, he took possession of it for the purpose of preventing another from committing or continuing to commit an offence in connection with that drug and that as soon as possible after taking possession of it he took all such steps as were reasonably open to him to destroy the drug or to deliver it into the custody of a person lawfully entitled to take custody of it; or

(b) that, knowing or suspecting it to be a controlled drug, he took possession of it for the purpose of delivering it into the custody of a person lawfully entitled to take custody of it and that as soon as possible after taking possession of it he took all such steps as were reasonably open to him to deliver it into the custody of such a person.

This defence would cover the situation where a parent finds their child in possession and, taking the drug from the child, flushes it down the toilet or hands it in at a police station.

Similarly, it will cover the person who finds a controlled drug in the street and is taking it to the police station to hand in.

We should now consider who may be in **lawful possession**.

The Misuse of Drugs Regulations 2001 define the persons who may have lawful possession.

The Secretary of State may issue a licence to possess a controlled drug, subject to the conditions of the licence.

Any of the following may lawfully possess a controlled drug:

(a) A constable acting in the course of his duty.

(b) A carrier acting in the course of that business.

(c) A post office employee acting in the course of that business.

(d) A person engaged in the work of a laboratory where the drug has been sent for forensic examination, when acting in the course of his duty.

(e) A person conveying the drug to a person authorised to have it in his possession.

The regulations also authorise certain medical personnel, masters of ships and managers of offshore installations to possess certain controlled drugs.

Defences (General)

Apart from those special defences relating solely to simple possession, there are a number of other defences which apply to all of the following offences under the 1971 Act:

Section 4(2) Unlawful Production

Section 4(3) Unlawful Supply

Section 5(2) Unlawful Possession

Section 5(3) Possession with intent to supply

Section 6(2) Unlawful Cultivation of Cannabis

Remember, the prosecution have to prove a case against the accused. However, the defendant can put forward evidence that they were not at fault by relying on one of the three defences within s. 28, as follows:

1. Lack of knowledge of a fact alleged

The 'fact' must be one which it is necessary for the prosecution to prove to get a conviction, ie one of the elements of the offence.

The accused could raise three things:

(a) They did not know the existence of 'some fact alleged'.

(b) They did not suspect the existence of that fact.

(c) They have no reason to suspect the existence of that fact.

2. Lack of knowledge of a controlled drug

(a) When the prosecution prove that the substance subject to the charge is a particular controlled drug, the defendant shall NOT be acquitted merely because they thought it was a different type of controlled drug.

HOWEVER

(b) If they prove that they neither believed nor suspected, nor had reason to suspect that the substance was a controlled drug, they WILL be acquitted.

3. Belief that it was a drug to which they were entitled

To avail themselves of this defence the accused must prove:

(a) they believed it was a controlled drug, or a particular controlled drug

AND

(b) if it had been that controlled drug or particular controlled drug, they would not have committed an offence.

Now look at the following circumstances and decide what defence, if any, could apply.

1. 'Steve' is unlawfully cultivating cannabis seedlings in his garden. 'He' goes on holiday and asks his neighbour to look after the garden for him. The neighbour waters the garden and generally tends the plants. 'Steve' does not tell the neighbour that the seedlings are cannabis and, as they are only immature seedlings, he has no idea what kind of plant they are.

2. 'Ali', a registered drug addict, goes to a chemist with a prescription for 5 ml of heroin, but the chemist makes a mistake and gives him 5 ml of pethidine.

3. A postal worker is delivering a brown paper wrapped parcel to a house, in the course of his duty. As she is taking it from her van she drops it and it splits open revealing a quantity of cannabis.

4. 'Jan', a known drug pusher, gives 'Jo' £200 to deliver a small package to an address half a mile away. The package is found to contain cannabis.

5. 'Luke' has a headache and comments on this. 'Sue' gives him tablets marked 'DICONAN' (which is a controlled drug) and says they will help. 'Luke' thinks they are a normal headache tablet like 'Aspro'.

Answers

1. Lack of knowledge of fact alleged, i.e. that they were cannabis plants.

2. Belief that it was a drug to which he was entitled, i.e. 'Ali' is not in unlawful possession of the pethidine but he believes it is heroin and, if it were, he would be in lawful possession.

3. Lack of knowledge of a controlled drug; no reason to suspect that what he was delivering was a controlled drug.

4. No defence—because 'Jo' had 'reason to suspect it contained a controlled drug' in the circumstances.

5. Lack of knowledge of a controlled drug, i.e. 'Sue' knows Luke has 'DICONAN', a drug, but Luke does not know it is a controlled drug.

Before we look at police powers, there are two further offences that demonstrate the wide ranging scope of this legislation.

Incitement

In our case study, Mack commits two offences of incitement (s. 19, Misuse of Drugs Act 1971) when he asks Oscar to supply him cocaine and when he gets Katie to commit offences of offering and being concerned in the supply of cocaine. Had either Oscar or Katie been an undercover officer Mack would still have committed incitement.

Assisting or Inducing an Offence outside the UK

Oscar, when he orders his drug assignment from Spain, induces others to commit similar drugs offences in Spain. Oscar commits this offence in this country (s. 20, Misuse of Drugs Act 1971).

Police Powers

Section 23, Misuse of Drugs Act 1971

1. Powers to stop, search and detain.

If a Constable has reasonable grounds to suspect that any person is in possession of a controlled drug, he may:-

(a) Search that person and detain him for the purpose of searching.

(b) Search any vehicle or vessel in which the constable suspects that the drug may be found.

(c) Seize and detain anything found in the course of the search which appears to be evidence of an offence under the Act.

The search must be conducted in accordance with the PACE Codes of Practice and the power to 'detain' extends to removal to a nearby Police Station when a more thorough search is necessary.

Search warrants

Warrants are also obtained under s. 23 by laying a sworn information before a Magistrate:

(a) Issued to a constable acting for the police area in which the premises are situated.

(b) Execution is by any constable at any time within one month of the date of issue.

(c) Authority is given to search the premises and any person therein.

(d) Any controlled drugs or documents relating to drugs transactions may be seized. (Note: could also remove other articles which could be evidence under PACE).

(e) Force may be used to enter premises, if necessary.

Obstruction

Section 23(4) also creates the offence of obstruction. A person commits an offence if he or she:

(a) intentionally obstructs a person in the exercise of his powers;

(b) conceals any such books, documents, stocks or drugs from a person acting in the execution of their duty; or

(c) without reasonable excuse fails to produce any such books or documents on the demand of a person in the exercise of these powers.

Powers of arrest

These are governed by PACE and as such relate to the punishments applicable to each offence. Now would be a good time to look at these and the best way to illustrate them is in the form of a table as follows:

SECTION	OFFENCE	DRUG CLASSIFICATION			PUNISHMENT
3	Unlawful Importation/Exportation	A	B		14 years
				C	5 years
4(2)	Unlawful Production	A	B		14 years
				C	5 years
6(2)	Unlawful Cultivation of Cannabis				14 years
4(3)	Unlawful Supply	A	B		14 years
				C	5 years
5(3)	Possession with Intent to Supply	A	B		14 years
				C	5 years
8	Controlled Drugs on Premises	A	B		14 years
				C	5 years
5(2)	Unlawful Possession	A			7 years
			B		5 years
				C	2 years
23(4)	Obstruction				2 years

Although cannabis is now a class C drug, possession of it has become an arrestable offence (schedule 1A PACE).

Drug Testing

Where:

• a person has been charged with a 'trigger offence'; or

• an inspector has reasonable grounds for suspecting that misuse of Class A drugs has caused or contributed to the offence charged;

• you can take a urine or non-intimate sample from the person (but not by force) to establish whether there is a Class A drug in their body.

'Trigger' offences include:
theft, robbery, burglary, aggravated burglary, taking a motor vehicle, aggravated taking of a motor vehicle, deception, handling, going equipped and ss. 4, 5(2), 5(3), Misuse of Drugs Act 1971 (Class A offences). Failing to supply a sample is a summary offence.

Summary

The offences relating to the misuse of drugs are many and are so wide-ranging that virtually any involvement with drugs, however minor, will fall within at least one of the offences. It is important for the investigator to be aware of the statutory defence as 'excuses' such as these will often be given.

Self-assessment Test

Having completed this Chapter, test yourself against the objectives outlined at the beginning.

1. Anouska buys some cannabis for use at a party. With regard to the Misuse of Drugs Act 1971, which class of controlled drugs, if any, does Anouska have in her possession?

   (a) Class A.
   (b) Class B.
   (c) Class C.
   (d) Cannabis is not classified as a controlled drug.

2. Sally knows where she can get some cocaine for herself. She said to her friend Liz, 'Do you want some cocaine this week when I get mine?' Liz replied, 'As long as its good stuff, get me enough for the weekend.' When Sally buys what she believes to be cocaine she is actually supplied with pure baking powder. She then unwittingly sells the baking powder to Liz for £50.

   With regard to the offences of possession and supplying, which of the following statements is correct?
   (a) When Sally bought the baking powder she committed the offence of possession with intent to supply cocaine.
   (b) When Sally asked Liz if she wanted some cocaine she committed the offence of offering to supply to Liz.
   (c) When Sally sold the baking powder to Liz she committed the offence of supplying cocaine.
   (d) When Sally bought the baking powder from her supplier she committed the offence of possession of cocaine.

3. Trevor had to report weekly to the police station whilst on bail for drugs offences. An hour before Trevor reported at the police station he gave his friend Lyle three wraps of heroin to look after so that Trevor wouldn't get caught with them when he reported. After reporting, Trevor met up with Lyle who gave the wraps back to Trevor.

   With regard to possession and supplying contrary to the Misuse of Drugs Act 1971, which of the following statements is correct?
   (a) Lyle supplies Trevor with the heroin when he handed the wraps back to him.
   (b) Trevor supplies Lyle when he handed him the three wraps of heroin.
   (c) When Lyle looks after the heroine he commits the offence of obstructing the supply of heroin.
   (d) When Lyle receives the three wraps he commits the offence of possession of heroin.

4. Eduardo persuaded his wife Lisa to buy a residential basement flat and to let it to young professional couples as an investment. He took control of the refurbishment and letting arrangements while Lisa agreed to look after the financial affairs of the flat. Once the flat was ready for letting, Eduardo found it impossible to find any tenants. In order that Lisa could pay the mortgage, Eduardo allowed an old friend to use the flat temporarily to make drugs using the kitchen as a laboratory. Eduardo thought his friend was going to make ecstasy for fun but in fact produced LSD for supplying. He lied to Lisa about his friend and told her that he had found some reliable professional tenants.

With regard to s. 8, Misuse of Drugs Act 1971, who, if anybody, commits the offence of being concerned in the management of premises for the production of a controlled drug?

(a) Only Lisa because she deals with all of the financial management of the flat.

(b) Both Lisa and Eduardo because they jointly manage the flat.

(c) Only Eduardo because he is the only one who knows what his friend is doing.

(d) Neither Lisa nor Eduardo because Eduardo thinks his friend is making ecstasy not LSD.

5. George was late for work and had run out of cigarettes. Knowing that his son kept a spare packet in his bedroom, he found the spare packet and put them in his pocket. Unbeknown to George, his son had rolled some reefer cigarettes using cannabis and hidden them in the spare packet.

   Which of the following statements is correct?

   (a) George is in possession of cannabis and commits the offence of possession with no defence available to him.

   (b) George is not in possession of the cannabis because he did not put the cigarettes in the packet.

   (c) George is in possession of the cannabis but would have a defence to the offence of possession.

   (d) George is not in possession of the cannabis because he does not know the reefers are in the packet.

6. With regard to the defences written into the Misuse of Drugs Act 1971, which of the following set of circumstances would **not** be covered by the general defences detailed in s. 28.

   (a) A registered drug addict picked up his prescription for heroin from the local chemist shop but had been given methadone by mistake through an administrative error.

   (b) A mother found some ecstasy tablets in her daughter's bedroom and flushed them down the toilet to make sure her daughter didn't use the tablets.

   (c) A young girl of 15 years was asked to deliver a sealed envelope to a local address and given £5 for doing so. The envelope contained cocaine but the girl thought she had delivered a message.

   (d) A young man moved into a furnished flat and started watering what he though were tomato plants on the window ledge. They were in fact cannabis plants.

7. Constable Prince went with other officers to execute a search warrant for controlled drugs at a large house on the edge of town. The officers entered the house and secured the scene. Kevin, the occupier, had a quantity of crack cocaine hidden in the house. Kevin also knew that the telephone numbers of his drugs contacts were written in his diary and would incriminate him. He immediately threw the diary onto the open coal fire and the diary went up in flames.

   With regard to your powers under s. 23, Misuse of Drugs Act 1971, which of the following statements is correct?

   (a) Kevin did not commit the offence of obstruction because obstruction does not apply to search warrants.

   (b) Kevin committed the offence of obstruction because he intentionally destroyed evidence of a drugs offence.

   (c) Kevin could not commit the offence of obstruction because any obstruction must involve a physical obstruction of the officers themselves from carrying out the search.

   (d) Kevin committed the offence of obstruction because he was legally obliged to surrender any evidence found in his possession.

Answers to Self-assessment Test

1. (c) Cannabis is a Class C drug.

2. (b) A drug does not have to be in existence for the offence of 'offering to supply' to be committed. Where a person is in possession of a substance that is not in fact a drug, they cannot be in possession of a controlled drug. However, these situations would probably be attempts to commit offences and covered by the Criminal Attempts Act 1981.

3. (d) When looking after a drug for another, the return of it is not a supply. However, when the person looking after drugs benefits from doing so, it will be a supply when the drugs are returned. Any possession before that return of the drugs will be 'unlawful possession'.

4. (c) The person concerned in the management must know or turn a blind eye to the drug dealing on their premises. It is not necessary for the person to know the exact nature of the drug being produced.

5. (c) To prove possession you need to prove that the person had possession of the packet of cigarettes and that the person knew the packet had something in it. In this case George clearly has possession of the reefer. It would then be up to George to prove on a balance of probabilities that he had a defence under s. 28.

6. (b) There are two defences for possession of controlled drugs under s. 5 of the Act. This is one of them. The other answers cover the three main areas of the general defence under s. 28.

7. (b) Obstruction does apply to search warrants and amounts to any action to obstruct the officers in the exercise of their powers, not purely physical obstruction of the officers themselves. Kevin is under no obligation under s. 23 to surrender evidence.

Firearms

Objectives

With regard to firearms, at the end of this section, from a given set of circumstances, you will be able to:

1. Identify a 'firearm' under s. 57, Firearms Act 1968.

2. Identify 'ammunition' under s. 57, Firearms Act 1968.

3. Identify a 'shotgun' under s. 1, Firearms Act 1968.

4. Identify 'prohibited weapon' and 'prohibited ammunition' under s. 5, Firearms Act 1968.

5. Identify the points to prove for the offence of possession of a firearm with intent to endanger life contrary to s. 16, Firearms Act 1968.

6. Identify the points to prove for the offence of possession of a firearm with intent to resist arrest contrary to s. 17(1), Firearms Act 1968.

7. Identify the points to prove for the offence of possession of a firearm when committing a Schedule 1 offence contrary to s. 17(2), Firearms Act 1968.

8. Identify what constitutes an 'imitation' firearm.

9. Identify the points to prove for the offence of carrying a firearm with intent to commit an indictable offence contrary to s. 18, Firearms Act 1968.

10. Distinguish between 'has with him' and 'possession'.

11. Identify the points to prove for the offence of carrying a firearm in a public place contrary to s. 19, Firearms Act 1968.

12. Identify the points to prove for the offences of trespassing with a firearm contrary to s. 20(1) and (2), Firearms Act 1968.

13. Identify the powers given to police under s. 47, Firearms Act 1968.

Introduction

The Firearms Act 1968 was introduced to provide a system of control on the manufacture, sale and possession of the more dangerous types of weapon, to cause records to be kept to aid in the detection of illegal dealings, and to discourage the use of firearms in the commission of criminal offences. This Act has subsequently been amended by the Firearms Acts of 1982 and 1988.

Firearms and Ammunition

What is a firearm?

From your previous training you should be able to remember the definition of a firearm. To see how much you can remember try the next activity.

Below is an incomplete definition of a firearm with spaces provided for the missing key words. Spend a few minutes completing the definition.

A Firearm means any .... weapon of any description from which any ....
.... or other .... can be discharged and includes:

(a) any .... weapon whether it is such a .... weapon as aforesaid or not; and

(b) any .... of such a .... or .... weapon; and

(c) any .... to such weapon designed or adapted to diminish the .... or ....
    caused by firing the weapon.

See if your answer is correct by looking at the definition.

Section 57, Firearms Act 1968

'Firearm' means any lethal barrelled weapon of any description from which any shot, bullet, or other missile can be discharged and includes:

(a) any prohibited weapon, whether it is such a lethal weapon as aforesaid or not: and

(b) any component part of such a lethal or prohibited weapon; and

(c) any accessory to any such weapon designed or adapted to diminish the noise or flash caused by firing the weapon.

What about ammunition?

The definition of ammunition also comes from s. 57, Firearms Act 1968. It is short and easy to understand.

Section 57, Firearms Act 1968

'Ammunition' means ammunition for any firearm and includes grenades, bombs and other like missiles whether capable of use with a firearm or not and also includes prohibited ammunition

Ammunition is anything that flies out of the barrel of a gun or any other exploding device whether or not it is fired from a gun.

Shotguns

How would you recognise a shotgun if you saw one?

If you can answer the question below correctly, you would be able to recognise one.

A shotgun is a ...

(a) rifled barrel gun

(b) smooth bore gun

(c) gun with a barrel not less than 24 inches in length

(d) gun with a barrel not less than 22 inches in length

... not being an air gun.

(i) (a) and (c)   (iii) (b) and (c)

(ii) (a) and (d)   (iv) (b) and (d)

Answer:...

Look at the definition below where you will see that (iii) correctly describes a shotgun.

A shotgun is a smooth bore gun (not being an air gun) with a barrel not less than 24 inches in length and a bore not exceeding 2 inches in diameter; and either has no magazine or has a non-detachable magazine or has a non-detachable magazine incapable of holding more than two cartridges; and is not a revolver gun.

It is an offence for a person to have in his possession or to purchase or to acquire a shotgun without holding a certificate.

Section 1 Firearms

Section 1 of the Firearms Act 1968 creates offences of possession, purchasing or acquiring a (Section 1) firearm or ammunition without holding a firearms certificate. Although not telling us what firearms it relates to, it does tell us what firearms are excluded.

Section 1 Firearm

... applies to every firearm except

(a) a SHOTGUN within the meaning of this Act, that is to say a smooth-bore gun (not being an airgun) which:
   (i) has a barrel not less that 24″ in length and does not have any barrel with a bore exceeding 2″ in diameter.
   (ii) either has no magazine or has a non-detachable magazine incapable of holding more than two cartridges; and
   (iii) is not a revolver gun.

(b) An AIR WEAPON that is to say an air rifle, air gun or air pistol not of a type declarded to be specially dangerous.

and

Section 1 Ammunition

... applies to any ammunition for a firearm except:

(a) Cartridges containing 5 or more shot, none of which exceed .36 inches in diameter.

(b) Ammunition for an air gun, air rifle, air pistol and

(c) Blank cartridges not more than one inch in diameter.

In the following table identify the types of firearm and ammunition.

Tick as many columns as appropriate to each weapon. We have not looked at prohibited weapons and ammunition yet but see if you can also identify which fit into those categories.

Description of of weapon	firearm	ammo	Section 1 f/arm	Section 1 ammo	Shotgun	Prohib weapon	Prohib ammo
Smooth bore gun with barrel length of 26″							
Cartridge containing 3 shot of .38 of an inch diameter							
air pistol which uses a self-contained gas cartridge system							
CS gas spray can							
bullet for .22 rifle							
plastic bullet (baton round)							
200 bullets for machine gun							
machine gun							
air rifle pellet							
pump action smooth bore gunbarrel length 20″							
smooth bore gun with barrel length 14″							
blank cartridge 2″ diameter							
automatic pistol disguised as a fountain pen							
revolver—overall length 10 cm							

Description of of weapon	firearm	ammo	Section 1 f/arm	Section 1 ammo	Shotgun	Prohib weapon	Prohib ammo
Smooth bore gun with barrel length of 26"	✓				✓		
Cartridge containing 3 shot of .38 of an inch diameter		✓		✓			
air pistol which uses a self-contained gas cartridge system	✓					✓	
CS gas spray can						✓	
bullet for .22 rifle		✓		✓			
plastic bullet (baton round)		✓					
200 bullets for machine gun		✓				✓	
machine gun	✓					✓	
air rifle pellet		✓					
pump action smooth bore gunbarrel length 20"	✓					✓	
smooth bore gun with barrel length 14"	✓		✓				
blank cartridge 2" diameter		✓		✓			
automatic pistol disguised as a fountain pen	✓					✓	
revolver—overall length 10 cm	✓					✓	

Were you able to identify the prohibited weapons and prohibited ammunition? See how they fit the definition given in the Act.

Prohibited Weapons

Section 5, Firearms Act 1968 as amended, Firearms (Amendment) Act 1988

1(a) Any firearm which is so designed or adapted that two or more missiles can be successively discharged without repeated pressure on the trigger.

(ab) Any self-loading or pump-action rifle other than one which is chambered for .22 rim-fire cartridges.

(abc) Any firearm which either has a barrel less than 30 centimetres in length or is less than 60 centimetres in length overall, other than an air weapon, a muzzle-loading gun or a firearm designed as signalling apparatus.

(ac) Any self-loading or pump-action smooth-bore gun which is not chambered for .22 rim-fire cartridges and either has a barrel less than 24 inches in length or is less than 40 inches in length overall.

(ad) Any smooth-bore revolver gun other than one which is chambered for 9 mm rim-fire cartridges or loaded at the muzzle end of each chamber.

(ae) Any Rocket launcher or mortars, (except those designed for fireworks, signal flames and safety lines).

(af) Any air rifle, air gun or air pistol which uses, or is designed or adapted for use with, a self-contained gas cartridge system.

(b) Any weapon of whatever description designed or adapted for the discharge of any noxious liquid, gas or other thing.

(c) Explosive bullets. Grenades, bombs, explosive rockets and shells (if capable of being used with any firearm).

1A(a) Any firearm which is designed as another object.

(b) Any ammunition containing or designed or adapted to contain any noxious thing.

The Secretary of State, subject to the approval of Parliament, may add to the list of prohibited weapons and ammunition any firearm which was not lawfully on sale in substantial numbers before 1988, and which appears among other things, to be specially dangerous, and the Firearms Acts (Amendment) Regulations 1992 made some additions to the list of prohibited weapons. These additions mostly referred to military type weapons such as rockets and missiles, however, they include any firearms disguised as other objects for example, walking stick shot guns, pen pistols etc, that are not antiques held as curiosities or ornaments.

You may be wondering why some of the answers have been given. Let's deal with prohibited weapons and ammunition first. The machine gun clearly falls into part 1(a) of the definition. The CS canister is 1(b); if the CS is contained in a cartridge for firing from a shotgun it would also be 1A(b). The automatic pistol is a prohibited weapon because it is disguised as another object (1A(a)). The revolver is a prohibited weapon because it has an overall length less than 60 centimetres. How did you fare with the sawn-off shotguns? The rule basically is if it's single or double barrelled without a magazine containing more than two cartridges it's a Section 1 firearm, but if it has a magazine and is self loading or pump action it falls under (ac) above and is a prohibited weapon.

What about the air pistol?

Remember that most air weapons will be 'lethal barrelled weapons', and as such come within the definition of a firearm. The Forensic Science Laboratory will be able to tell you whether it is lethal or not. Those air weapons that propel their ammunition with a self-contained gas cartridge are now prohibited weapons.

You may be wondering why some of the weapons have a tick in more than one column. This is because the description fits two or more definitions. For instance, a shot gun is defined as such for certification purposes but is never-the-less still a firearm.

Let's recap on what we have covered so far

A Firearms Certificate is required to possess, purchase or acquire a Section 1 Firearm. A Section 1 Firearm is any firearm other than those exceptions previously outlined, ie a shotgun, an air weapon not of a type declared specially dangerous and ammunition for those types of firearm.

In addition we have discussed that shotguns similarly cannot lawfully be possessed, purchased or acquired without holding a shotgun certificate. So only those other firearms outlined (Air Weapons and ammunition for such weapons) may be possessed, purchased or acquired without some form of certificate.

Prohibited weapons are as their title suggests prohibited and may only be possessed, purchased, acquired, manufactured, sold or transferred with the authority of the Secretary of State.

Criminal Use of Firearms

The Firearms Act 1968 also created a series of offences relating to the misuse of firearms.

From your previous knowledge identify what offences contrary to firearms legislation, if any, have been committed in the following circumstances. If you need any help, you will find a useful table of offences following this activity.

Circumstances	Offence(s) committed
1. Keith went on a date with a work colleague. At the end of the evening he went back to her flat and had sexual intercourse without her consent. The work colleague reported the alleged rape to the police. Keith was arrested at his home address on suspicion of rape. At his home, when he was arrested, he had a legally held shotgun he used for clay pigeon shooting.	
2. Mohammed had a long-running feud with his neighbour over a strip of land that prevented Mohammed building a garage at the bottom of his garden. Mohammed became so frustrated that he borrowed an imitation handgun from his cousin and hid the handgun in his desk at work in the centre of town. He went to his neighbour and said 'Unless you let me build my garage I have a gun and will harm one of your children'. Mohammed knew that the gun was an imitation but he intended to make his neighbour believe that he would seriously harm one of the children with the gun if the neighbour refused to co-operate. His neighbour ignored the threat and was not concerned.	
3. Grant and two other men followed a security van in a stolen car. The van was delivering cash to banks in the town centre. When the van pulled into the rear private car park of one bank, Grant and the men followed. In the car park they waited for the security guard to get out of his cab. Grant and one of the men then jumped out of the car and Grant pointed a loaded sawn-off shotgun into the face of the guard and said 'Open the van and throw out the money or else you're dead'. The guard did as he was told and the three made off with the money. Both of the other men knew Grant had the loaded shotgun to carry out the robbery.	

Circumstances	Offence(s) committed
4. Marie had discovered that her partner David was having a long-term relationship with Lauren who lived in Spain. David legally kept a shotgun in a locked gun cabinet to which only he had access. Marie intended to go out to Spain and shoot Lauren with the shotgun if she refused to stop seeing David. Consequently, Marie tricked a local locksmith into opening the gun cabinet so that she could get ready access to the shotgun should she need it to carry out her threat.	
5. Moore went into an off-licence, forced the manager into the off-licence back office and robbed the manager of the day's takings. He had an imitation gun in his pocket to use to threaten the manager but did not need it. After the robbery the manager followed Moore out of the off-licence and shouted, 'Stop that man he has just robbed me!'. A young man rushed at Moore and grabbed him around the neck to stop him getting away. Moore reached into pocket and pulled out the gun and hit the young man in the face with it in order to get away. The young man held on until the police arrived. However, Moore was eventually acquitted of the robbery.	

Table of Criminal Use of Firearm Offences

It is an offence ...	Section	Characteristics of offence
For a person to have in their possession any firearm or ammunition with intent by means thereof to endanger life or to enable another person by means thereof to endanger life whether any injury is caused or not.	16	'Possession'—firearm need not be shown. Excludes imitation. Specific intent to endanger. . . . but does not have to be immediate and may be conditional. Life of other not self. Other may be outside UK. Possible self-defence (rare).
For a person to have in their possession any firearm or imitation with intent (a) by means thereof to cause, or (b) to enable another by means thereof to cause, any person to believe that unlawful violence will be used against them or another person.	16A	'Possession'—firearm need not be shown. Includes imitation. Specific intent to cause ....
For a person to make or attempt to make use whatsoever of a firearm or imitation with intent to resist or prevent the lawful arrest or detention of himself or another person.	17(1)	'Actual use' or attempted 'actual use'. Excludes component parts. Includes imitation but not imitation parts. Specific intent to resist .... Arrest must be lawful.
At the time of them committing or being arrested for a Schedule 1 offence, has in their possession a firearm or imitation, unless they can show they had it in their possession for a lawful object.	17(2)	'Possession' committing offence or 'possession' being arrested for offence. Includes imitation. Schedule 1 offences only (DART). No need for conviction for DART offence.

It is an offence ...	Section	Characteristics of offence
For a person to have with them a firearm or imitation with intent to commit an indictable offence, or resist or prevent the arrest of another, in either case while they have a firearm or imitation with them.	18	'Has with them'. Includes imitation. Specific intent to commit indictable . . . (not to use firearm) Indictable (includes either way offences). Arrest need not be lawful.
If, without lawful authority or reasonable excuse (the proof whereof lies on them), they have with them in public place loaded shotgun or an air weapon (whether loaded or not), any other firearm (whether loaded or not) together with ammunition suitable for use in that firearm, or an imitation firearm.	19	'Has with them'. Includes imitation. Usual public place definition. Absolute offence. Loaded for shotgun. Loaded or unloaded for air weapon. Loaded or unloaded for any other firearm but with ammunition.
If, while they have a firearm or imitation with them, they enter or are in any building or part of a building as a trespasser and without reasonable excuse (the proof whereof lies on them).	20(1)	'Has with them'. Includes imitation. No need to enter with firearm.
If, while they have a firearm or imitation with them, they enter or are on any land as a trespasser and without reasonable excuse (the proof whereof lies on them).	20(2)	'Has with them'. Includes imitation. No need to enter with firearm. Includes land covered by water.

These definitions should help you distinguish between the different offences.

'Possession'	'Constructive' possession give wide interpretation meaning having firearm under their control. Does not need specific knowledge of firearm as long as knowledge of possession of something.
'Has with them'	Narrow interpretation meaning a close physical link and a degree of immediate control (ie readily accessible as on their person or in a car nearby but not in a house a few miles away).
'Imitation firearm'	For criminal use of firearms offences (ss. 16–20) means in the general sense as those having the appearance of a firearm (it does not have to be readily converted into a s. 1 firearm). Where someone holds their fingers inside their jacket, in circumstances where they clearly intend a victim to believe they have a firearm and threatening to shoot them, the fingers amount to an imitation firearm.
'Schedule 1 offence'	DART—Damage (s. 1, 1968 Act), Assaults (ss. 47, 20–22, 38, 1861 Act), Rape (ss. 1, 17, 18, 20, 1956 Act), Theft (ss. 1, 8, 9, 12, 21, 1968 Act).
'Loaded'	Means if there is ammunition in the chamber or barrel (or in any magazine or other device) whereby the ammunition can be fed into the chamber or barrel by the manual or automatic operation of some part of the weapon.

Here are the answers to the problems posed:

1. Keith committed the offence under s. 17(2). He has probably committed both parts of the offence because he probably also had possession of the shotgun when he committed the rape (Schedule 1 offence). He definitely committed the offence when he was arrested for a Schedule 1 offence. However, Keith would appear to have a defence of having possession of the shotgun with a lawful purpose (for him to prove).

2. Mohammed did not commit the offence under s. 16A because although Mohammed intended to cause his neighbour to believe that he would injure the neighbour's children, the neighbour needs to know that the gun exists. Had the neighbour known Mohammed had the firearm, possession of the imitation by Mohammed would be enough for the offence under s. 16A.

3. Grant and the two other men committed offences under ss. 17(2), 18, 19 and 20(2). All three have possession of the loaded firearm because they know Grant has it in his possession. When in the car following the van they committed the offence under s. 19. In the private car park they committed the offence under s. 20(2). When they committed the robbery they also committed offences under ss. 17(2) and 18. For the ss. 18, 19 and 20(2) offences the three need to have the firearm with them, which they do in this instance.

4. Marie committed the offence under s. 16. When she tricked the locksmith she gained possession of the shotgun. She intended to injure Lauren conditionally ie that she give up the relationship. If she carries out the threat with a shotgun this is likely to endanger her life. Provided she has the intention and possession of the firearm in this country the offence is complete even though the victim is abroad. Lauren does not need to be injured.

5. Moore committed offences under ss. 17(2), 18, 19, 20(1) and 17(1). He committed the robbery and had the imitation firearm with him and in his possession. He then trespassed in the off-licence office with the imitation firearm. Finally, Moore made use of the imitation firearm to resist the arrest by the young man and can be found guilty even though he was eventually acquitted of the robbery. It is an offence to have an imitation firearm in a public place.

Police powers: Section 47, Firearms Act 1968

It is an offence for a person having a firearm or ammunition with him to fail to hand it over when required to do so by a constable.

Power to stop and search: Section 47(1) and (3)

A constable may require any person whom he has reasonable cause to suspect:

a) of having a firearm with or without ammunition with him in a public place; or

b) to be committing or about to commit, elsewhere than in a public place, an offence relevant for the purpose of this section (*under Sec 18 or 20),

to:- hand over the firearm or any ammunition for examination by the constable, and the constable may search that person and may detain him for the purpose of doing so.

Power to stop and search vehicles: Section 47(4)

If a constable has reasonable cause to suspect that there is a firearm in a vehicle in a public place, or that a vehicle is or is about to be used in connection with the commission of an offence relevant for the purpose of this section (*Section 18 and 20) elsewhere than in a public place, he may search the vehicle and for that purpose require the person driving or in control of it to stop it.

Power of entry: Section 47(5)

For the purpose of exercising the powers conferred by Section 47 a constable may enter any place.

* You may well be wondering why this power is expressly given to the less serious offences of ss. 18 and 20. If you look at ss. 16 and 17 you will see that they contain a section 18 offence. (This may be compared with Robbery as in every Robbery there is a theft).

Production of Certificates

Production of Certificates: Section 48(1)

A constable may demand from any person whom he believes to be in possession of a firearm or ammunition to which Section 1 of this act applies, or shotgun, the production of a certificate.

Failure to produce a Certificate: Section 48(2) and (3)

If a person upon whom a demand is made fails to produce the certificate, or permit the constable to read it or show he is entitled to have possession of the firearm, ammunition or shotgun without holding a certificate, the constable may seize and retain the firearm, ammunition or shotgun and require his name and address.

Refusal or failing to give true name and address is an offence.

Firearms Act 1982

This Act extends the provision of s. 1 of the Firearms Act 1968 to include certain imitation firearms.

A person commits an offence if without holding a valid firearms certificate or otherwise than as authorised by such a certificate, he has in his possession, or purchases or acquires, an imitation firearm as outlined:

Imitation Firearms requiring a Firearms Certificate

(a) It has the appearance of being a firearm to which Section 1 of the Firearms Act 1968 (firearms requiring a certificate) applies:

AND

(b) It is so constructed or adapted as to be readily convertible into a firearm to which section 1 applies.

Briefly, 'readily convertible' means that it may be converted without special skills or specialised tools.

Summary

Firearms legislation can appear quite complicated and in this section we have looked at the aspect you are most likely to become involved in.

Self-assessment Test

1. You are investigating a robbery where the suspect held a knife to the throat of the victim. You get reliable information from a registered source that John committed the robbery. You go to John's flat and arrest him for the robbery. You search the flat and find a sawn-off shotgun under his bed. John is later acquitted of the robbery at court.

With regard to possessing a firearm when committing a Schedule 1 offence, s. 17(2), Firearms Act 1968, which of the following statements is correct?

(a) John cannot be guilty of this offence because he has to be in possession of the firearm he used in the robbery when he is arrested.

(b) John cannot be guilty of this offence because he was later acquitted of the robbery.

(c) John is guilty of this offence because he is in possession of a firearm when being arrested for a Schedule 1 offence.

(d) John is guilty of this offence because he is in possession of a weapon when committing a Schedule 1 offence.

2. Alex was a small-time heroin dealer on the local estate. Alex and his partner Theresa went to deal from the park toilets, keeping wraps of heroin in the boot of their car nearby. They also kept an imitation handgun in the boot to scare anyone who got awkward with them. Local police officers had been keeping the two under observation and moved in to arrest Alex. As they were arresting Alex in the park, Theresa jumped on the back of the arresting officer in an attempt to allow Alex to escape.

With regard to having a firearm with intent to commit an indictable offence contrary to s. 18, Firearms Act 1968, which of the following statements is correct?

(a) Theresa commits the offence of having an imitation firearm with her when she intends to prevent Alex's arrest.

(b) Alex commits the offence of having an imitation firearm with him intending to resist arrest.

(c) Alex cannot commit the offence because he did not intend to use the imitation firearm in an indictable offence.

(d) Theresa cannot commit the offence because the imitation firearm is in the boot of the car when she intends to prevent Alex's arrest.

3. Robert had built up a massive gambling debt at an illegal gaming club. He stole some of the money he owed and took it to the owner of the club. The club owner had an imitation handgun in his safe at the club. When Robert arrived at the club, the owner showed Robert the handgun and said to him, 'I want the rest by Friday or else you will end up in the foundations of the building site opposite with a bullet in your head'.

With regard to possession of a firearm with intent to endanger life or cause fear of violence, s. 16 and 16A, Firearms Act 1968, which of the following statements is correct?

(a) The club owner committed the offence of possession of a firearm with intent to endanger life.

(b) The club owner committed the offence of possession of an imitation firearm with intent to cause fear of violence.

(c) The club owner cannot commit the offence of possession of a firearm with intent to endanger life because he did not have the gun in his physical possession.

(d) The club owner cannot commit the offence of possession of an imitation firearm with intent to cause fear of violence because Robert could not be frightened.

4. Abu went to a party, as a guest of an art critic, for the opening of an art exhibition in the public rooms of the art gallery. After a short while he decided to go into the private offices of the gallery. In a cupboard he found an unloaded air rifle. He brought it out of the cupboard and put it in a position so that should he be disturbed he could quickly pick it up.

With regard to s. 20(1), Firearms Act 1968, has Abu committed the offence of trespassing with a firearm?

(a) No, because he had the firearm with him after he trespassed in the gallery office.

(b) No, because the air rifle is not a firearm for this offence.

(c) Yes, because you do not have to enter the office as a trespasser with the firearm.

(d) Yes, but only because the air rifle was loaded.

5. Neil broke into a house when the occupiers were away on holiday. He knew that there was a shotgun locked in a gun cabinet and he wanted it for his own use. He broke open the gun cabinet and stole the shotgun, but he could not find any suitable ammunition to go with the gun. Neil left the house and drove to a shooting club where he knew he could steal some ammunition for the shotgun. After he had stolen 100 rounds of ammunition suitable for the shotgun he drove to a local pub for a drink. He left the shotgun and ammunition locked in the boot of his car.

With regard to the Firearms Act 1968, which of the following statements is correct?

(a) Neil commits no criminal use of firearms offences.

(b) Neil commits the offence of possession of a firearm together with suitable ammunition in a public place contrary to s. 19.

(c) The car park of the pub is not a public place for offences contrary to s. 19.

(d) Neil commits the offence of being in possession of a firearm when committing the burglaries contrary to s. 17(2).

Answers to Self-assessment Test

1. (c) John does not have to commit the Schedule 1 offence with a firearm or imitation firearm when committing the offence of being in possession of a firearm whilst being arrested for such an offence. The original Schedule 1 offence does not have to result in a conviction. This section deals with firearms or imitation firearms, not other weapons.

2. (a) In the question there is no evidence that Alex intends to resist arrest. Alex could be committing the offence of having the imitation firearm with him whilst committing an indictable offence (possessing controlled drugs with intent to supply). He does not have to intend to use the imitation firearm in order to commit the indictable offence. Theresa does have the imitation firearm with her because it is readily accessible when it is in the car boot.

3. (b) An imitation firearm is not included in the offence of possession with intent to endanger life, but it is included for s. 16A. You would only need to prove the owner had possession and intended to cause Robert to fear he would be subjected to violence by use of the firearm.

4. (c) This offence is different from aggravated burglary where the person enters a building or part of a building as a trespasser and has a weapon of offence with them. Under s. 20(1) Abu can enter the office as a trespasser with the air rifle or simply be present in the office as a trespasser with the air rifle. An air weapon is a firearm and does not need to be loaded for this offence. Abu has the firearm with him because it is easily accessible and under his immediate control.

5. (d) Neil commits several criminal use of firearms offences, but only s. 17(2) is a valid option provided in the question. Burglary is a Schedule 1 offence and Neil commits to burglaries whilst in possession of the firearm. The shotgun has to be loaded for an offence under s. 19 and the pub car park would be a public place for the offence.

PART THREE

Sexual Offences

The Sexual Offences Act 2003

Objectives

With regard to the Sexual Offences Act 2003, at the end of this section, from a given set of circumstances, you will be able to:

1. Identify what is meant by 'sexual' (s. 78, Sexual Offences Act 2003).
2. Identify what is meant by 'touching' (s. 79(8), Sexual Offences Act 2003).
3. Identify the points to prove for the offence of rape (s. 1, Sexual Offences Act 2003).
4. Identify the points to prove for assault by penetration (s. 2, Sexual Offences Act 2003).
5. Identify the points to prove for assault by touching (s. 3, Sexual Offences Act 2003).
6. Identify the points to prove for causing sexual activity without consent (s. 4, Sexual Offences Act 2003).
7. Identify what is meant by 'true consent' (s. 74, Sexual Offences Act 2003).
8. Distinguish between 'evidential' and 'conclusive presumptions' in consent (ss. 75 and 76, Sexual Offences Act 2003).
9. Identify the points to prove for the offence of administering a substance with intent (s. 61, Sexual Offences Act 2003).
10. Identify the points to prove for the offence of committing a criminal offence with intent to commit a sexual offence (s. 62, Sexual Offences Act 2003).
11. Identify the points to prove for the offence of trespass with intent to commit a sexual offence (s. 63, Sexual Offences Act 2003).

Introduction

The concepts of indecency and indecent assault in previous law have been brought up to date by the Sexual Offences Act 2003. The Act introduces the terms 'sexual' and 'touching'. Consent continues to be a significant feature of the offences in the Act but has now been clearly defined. Look first at the definition of 'sexual' in the Act.

Sexual

Section 78 of the Act states that penetration, touching or any other activity will be sexual if a reasonable person would consider that:
(a) whatever the circumstances or any person's purpose in relation to it, it is sexual by its very nature or,
(b) because of its nature it may be sexual and because of its circumstances or the purpose of any person in relation to it, it is sexual.

Look at the following circumstances and test your understanding of 'sexual'. Tick the box where you think a reasonable person would consider the activity to be 'sexual' or not.

Circumstances	'Sexual' by its very nature	May be 'sexual' and because of its circumstances it is 'sexual'	May be 'sexual' and because of the purpose of any person it is 'sexual'	Not 'sexual' at all
1. A nurse carrying out a breast examination of a young woman for medical reasons.				
2. A woman masturbating a man in a car park.				
3. A male shop assistant putting a shoe on the foot of a female customer to get sexual gratification.				
4. A doctor carrying out a vaginal examination for a completely unrelated complaint to get sexual gratification.				

You may have considered circumstances:

1. Not 'sexual' at all, because a reasonable person would consider the examination to be purely a medical matter.

2. To be 'sexual' by its very nature because a reasonable person would recognise masturbation to be a sexual act in its own right.

3. Not to be sexual at all, because although it may appear to be sexual, it is unlikely that a reasonable person would consider such a fetish to be 'sexual'.

4. To be sexual, because the examination may be sexual and both the purpose of the doctor and its circumstances make it 'sexual'.

Section 79(8) provides the broad definition of 'touching' as follows.

Touching includes touching:

• with any part of the body

• with anything else

• through anything

and in particular, touching amounting to penetration.

Rape and Sexual Assaults

There are four sexual offences with similar features in the Act. Before we deal with consent, look at the four offences.

Rape (s. 1, Sexual Offences Act 2003)

A person (A) commits an offence if—

(a) he intentionally penetrates the vagina, anus or mouth of another person (B) with his penis,

(b) B does not consent to the penetration, and

(c) A does not reasonably believe that B consents.

Assault by penetration (s. 2, Sexual Offences Act 2003)

A person (A) commits an offence if—

(a) he intentionally penetrates the vagina or anus of another person (B) with part of his body or anything else,

(b) the penetration is sexual,

(c) B does not consent to the penetration, and

(d) A does not reasonably believe that B consents.

Sexual assault by touching (s. 3, Sexual Offences Act 2003)

A person (A) commits an offence if—

(a) he intentionally touches another person (B),

(b) the touching is sexual,

(c) B does not consent to the touching, and

(d) A does not reasonably believe that B consents.

Causing sexual activity without consent (s. 4, Sexual Offences Act 2003)

A person (A) commits an offence if—

(a) he intentionally causes another person (B) to engage in an activity,

(b) the activity is sexual,

(c) B does not consent to engaging in the activity, and

(d) A does not reasonably believe that B consents.

Now look at the following circumstances and test your understanding of these four offences. For the purposes of this exercise assume that B does not consent and A does not reasonably believe that B consents. Tick the box against the appropriate offence or indicate where no offence has been committed.

Circumstances	Section 1 Rape	Section 2 Assault by penetration	Section 3 Sexual touching	Section 4 Causing to engage in sexual activity	No offence
1. (A) deliberately places his hand over the breast of (B) in a sexual way.					
2. (A) deliberately places her fingers into the vagina of (B) in a sexual way.					
3. (A) accidentally presses his penis against the leg of (B) while standing on a crowded train.					

Circumstances	Section 1 Rape	Section 2 Assault by penetration	Section 3 Sexual touching	Section 4 Causing to engage in sexual activity	No offence
4. (A) deliberately inserts his penis into the mouth of (B), who is a male friend. (A) withdrew his penis before ejaculation.					
5. (A) deliberately inserts a vibrator sex toy into the mouth of (B) in a sexual way.					
6. (A) deliberately gets (B) to masturbate her male friend.					
7. (A) deliberately gets (B) to masturbate themselves.					

You should have identified that the following offences are committed in the circumstances described:

1. Sexual assault by touching. This offence can be committed by a man or woman on a man or woman.

2. Assault by penetration. This offence can be committed by a man or woman on a man or woman and can be penetration by a part of the body or anything else. Under s. 79(3) 'part of the body' includes references to a part surgically constructed (in particular, through gender reassignment).

3. No offence, because there must be an 'intention' by A to commit the offence. Had the touching been deliberate, the offence includes touching through clothing.

4. Rape. The mouth is one of the three orifices specified in this section (vagina and anus being the others). The victim of rape can either be a man or a woman. Rape can only be committed by a man but a woman can aid, abet, counsel or procure rape (e.g. holding down a victim who is being raped). The presence of semen or sperm would help prove penetration but ejaculation is not necessary for the offence of rape.

5. Assault by touching. Although assault by penetration can be committed by way of a part of the body or anything else (e.g. vibrator sex toy), the offence only applies to penetration of the vagina or anus. These circumstances amount to a sexual assault by touching any part of the body with anything else.

6. Causing a person to engage in sexual activity without consent. The offence can involve a number of permutations (e.g. a woman making a man penetrate her etc.). The offence may also involve just the victim engaging in a sexual act (see 7 below). Other people consenting in the offence with (A) may be aiding, abetting, counselling or procuring (A) to commit s. 4.

7. Causing a person to engage in sexual activity without consent.

Consent

Now let's look at the issue of consent in these sexual offences. Any consent must be 'true' consent, which is defined in the Act.

Section 74 states:

A person consents if he or she agrees by choice and has the freedom and capacity to make that choice.

The issue of consent is a question of fact. Because the act of penetration is a 'continuing' one from entry to withdrawal (s. 79(2)), consent can be withdrawn at any time during sexual activity. Continuing the penetration after consent has ended would amount to rape or assault by penetration.

If the victim is a child under 13, consent is not an issue because they cannot consent to ss. 1 to 4. There are specific offences when the victim is under 13 (s. 5 rape, s. 6 assault by penetration, s. 7 assault by touching and s. 8 causing or inciting a child under 13).

Presumptions and Consent

There are two presumptions that a court can consider when deciding whether consent was freely given.

1. Evidential Presumptions

This presumption accepts that consent was not freely given where certain circumstances exist. However, with evidential presumptions the defendant is allowed to put forward evidence in their defence to rebut this presumption. Look at the presumption as it appears in s. 75 of the Act:

(1) If in proceedings for an offence to which this section applies it is proved that—
  (a) the defendant did the 'relevant act'
  (b) any of the circumstances specified in subsection (2) below existed, and
  (c) the defendant knew that those circumstances existed.

the complainant is to be taken not to have consented to the relevant act unless sufficient evidence is adduced to raise an issue as to whether he consented, and the defendant is to be taken not to have reasonably believed that the complainant consented unless sufficient evidence is adduced to raise an issue as to whether he reasonably believed it.

This effectively means that where it is proved that the 'relevant act' (i.e. the penetration, touching, causing them to engage in sexual activity) occurred and circumstances of violence, restraint, drugs use etc. existed, and the defendant knew that those circumstances existed, it is accepted by the court that the complainant did *not* consent to the 'relevant act'. It is then up to the defendant to rebut the presumption by providing evidence to satisfy the court that the complainant was consenting and the defendant reasonably believed they were consenting.

The circumstances detailed in the Act are that:

(a) any person was, at the time of the relevant act (or immediately before it began), using violence against the complainant or causing the complainant to fear that immediate violence would be used against him/her;

(b) any person was, at the time of the relevant act (or immediately before it began), causing the complainant to fear that immediate violence was being used, or that immediate violence would be used, against another person;

(c) the complainant was, and the defendant was not, unlawfully detained at the time of the relevant act;

(d) the complainant was asleep or otherwise unconscious at the time of the relevant act;

(e) because of the complainant's physical disability, the complainant would not have been able at the time of the relevant act to communicate to the defendant whether the complainant consented;

(f) any person had administered to or caused to be taken by the complainant, without the complainant's consent, a substance, which having regard to when it was administered or taken, was capable of causing or enabling the complainant to be stupefied or overpowered at the time of the relevant act.

2. Conclusive Presumption

There are two types of situation where it will always be accepted that consent was not freely given and the defendant knew this. In these cases the defendant is not allowed to put forward evidence in their defence to rebut this presumption (it is conclusive). Look at the presumption as it appears in s. 76 of the Act:

> If it is proved that the defendant did the relevant act and that he/she:
>
> (a) intentionally deceived the complainant as to the nature or purpose of the relevant act, or
> (b) intentionally induced the complainant to consent to the relevant act by impersonating a person known personally to the complainant there will be a conclusive presumption both that the victim did not consent and also that the defendant did not believe she or he consented.

This effectively means once it has been proved beyond reasonable doubt that one of these circumstances existed, the court must assume that the complainant did not consent and the defendant did not believe he or she consented.

Whether a belief in consent is reasonable is to be determined having regard to all the circumstances, including any steps A has taken to ascertain whether B consents. It will be for the jury to decide whether any of the particular attributes of the defendant, such as extreme youth or disability, are relevant to their deliberations, subject to directions from the judge where necessary.

Now look at the following circumstances and test your understanding of the two presumptions under ss. 75 and 76. For the purposes of this exercise assume that it has been proved that A did the relevant act. Tick the box against the appropriate presumption or indicate where no presumption applies.

Circumstances	Evidential presumption (s. 75) applies	Conclusive presumption (s. 76) applies	No presumption applies	If s. 75 or 76 applies, is A allowed to rebut it?
1. (A) strikes (B) in the face immediately before inserting a finger into her vagina.				
2. (A) and (B) are held hostage by a terrorist and (A) rapes (B).				

Circumstances	Evidential presumption (s. 75) applies	Conclusive presumption (s. 76) applies	No presumption applies	If s. 75 or 76 applies, is A allowed to rebut it?
3. (A) intentionally deceives (B) about improving her voice by placing his penis in the mouth of (B).				
4. (A) intentionally deceives (B) into consenting to masturbate him by getting into bed with her in the middle of the night and pretending to be B's boyfriend.				
5. (A) crept into the bedroom of (B) while he was asleep and inserted his finger into the anus of (B).				
6. (A) put a stimulant drug into the drink of (B) to make her more sexually aroused and alert. Then (A) inserted his penis into her mouth.				

You should have identified that the following were applicable in the circumstances described:

1. Evidential presumption (using violence) allowed to be rebutted by A.

2. No evidential presumption can be made because only B to be unlawfully detained.

3. Conclusive presumption (intentionally deceiving the complainant as to the nature or purpose) and cannot be rebutted by A.

4. Conclusive presumption (intentionally inducing the complainant to consent to engaging in sexual activity by impersonating her boyfriend) and cannot be rebutted by A.

5. Evidential presumption (complainant was asleep) allowed to be rebutted by A.

6. No evidential presumption can be made because s. 75 only includes a substance which stupefies or overpowers the complainant.

Preparatory Offences

There are three specific offences that can be committed before a sexual offence takes place.

Administering a substance with intent (s. 61, Sexual Offences Act 2003)

(1) A person commits an offence if he intentionally administers a substance to, or causes a substance to be taken by, another person (B):

    (a) knowing that B does not consent, and

    (b) with the intention of stupefying or overpowering B, so as to enable any person to engage in a sexual activity that involves B.

This offence includes the wide intention to engage in sexual activity. Sexual activity could include stripping someone or taking pornographic photos of them. The offence will be committed at the stage of administering the substance, so no actual sexual activity needs to occur.

Committing an offence with intent to commit a sexual offence (s. 62, Sexual Offences Act 2003)

(1) A person commits an offence under this section if he commits any offence with the intention of committing a relevant sexual offence.

(2) In this section, 'relevant sexual offence' means any offence under this Part (including an offence of aiding, abetting, counselling or procuring such an offence).

This offence would include where a person subjects another person to violence or detains them, with the intention of committing a sexual offence (Part 1 covers sections 1 to 79 which basically includes all the sexual offences involving a victim of some sort).

Trespass with intent to commit a sexual offence (s. 63, Sexual Offences Act 2003)

(1) A person commits an offence if:

    (a) he is a trespasser on any premises,

    (b) he intends to commit a relevant sexual offence on the premises, and

    (c) he knows that, or is reckless as to whether, he is a trespasser.

(2) In this section:
'premises' includes a structure or part of a structure;
'relevant sexual offence' has the same meaning as in section 62;
'structure' includes a tent, vehicle or vessel or other temporary or movable structure.

This offence has replaced the burglary offence under s. 9(1)(a) Theft Act 1968 where rape was the ulterior offence. This new offence has the effect of widening the legislation to include trespass with intent to commit *any* of the sexual offences in the Act (Part 1) not just rape.

Summary

There are now four main offences that rationalise the previous legislation and common law. Because consent is central to this area of the law, the Act sets out clearly, for the first time, a clear definition. Now, if a defendant in court wants to claim they believed the other person was consenting, they will have to show they have reasonable grounds for that belief. Until now, if a defendant could prove they honestly believed consent had been given, however unreasonable their belief was, they would have been acquitted. The Act recognises the injustice in that approach.

Self-assessment Test

1. George invited his niece into his bedroom and placed a vibrator sex toy on the outside of her vagina because it gave him sexual pleasure.

With regard to the term 'sexual' in s. 78 Sexual Offences Act 2003, which of the following part, if any, of s. 78 applies?

(a) Whatever its circumstances, using the vibrator is 'sexual' by its very nature.

(b) Because of its nature, using the vibrator may be 'sexual' and only because of the circumstances is it 'sexual'.

(c) Because of its nature, using the vibrator may be 'sexual' and only because George is seeking sexual pleasure is it 'sexual'.

(d) Using the vibrator is not 'sexual' at all.

2. Ralph hid in some bushes alongside a footpath linking the railway station to a student hostel. When a young female student passed by, Ralph quickly ran up behind her and dragged her back into the bushes. He said to the student 'Keep quiet and you won't get hurt'. The student said 'Please don't do this'. He then intentionally placed his penis in the student's mouth but withdrew at the last moment and ejaculated over the ground.

With regard to ss. 1 to 4, Sexual Offences Act 2003, which of the following statements is correct?

(a) Ralph did not commit rape because he did not ejaculate inside the victim's mouth.

(b) Ralph did not commit rape because the offence is only committed by penetration of the vagina or anus.

(c) Ralph committed the offence of assault by penetration.

(d) Ralph committed the offence of rape because the student did not consent to the penetration and he did not reasonably believe that she consented.

3. Jack invited a friend back to his house after a long drinking session. Jack told his friend that his wife Anne liked group sex and when doing so liked to resist and protest as if she was not consenting. This was not true, but forcing his wife to have sex with a stranger gave Jack a feeling of power over his wife. The two men entered Jack's bedroom and found Anne awake in bed. Jack held his wife down while his friend intentionally penetrated her vagina with his penis. Anne continually resisted the sexual activity shouting 'No, get this man away from me'. The friend continued to penetrate Anne's vagina.

With regard to consent in s. 1, Sexual Offences Act 2003, which of the following statements is correct?

(a) There is a conclusive presumption in these circumstances that Jack's friend has intentionally induced Anne to consent by impersonating a person known personally to her.

(b) There is a conclusive presumption in these circumstances that Anne has been intentionally deceived as to the nature of the sexual activity.

(c) There is an evidential presumption in these circumstances that Anne did not consent and the friend can rebut this presumption by adducing sufficient evidence.

(d) There is an evidential presumption in these circumstances that Anne did not consent and the friend will be able to rebut it where he had an honest belief she was consenting.

4. Beth was very good friends with Ahmed, who was a male work colleague. They regularly went out after work for meals and drinks with other work colleagues. One evening after an office party Beth invited Ahmed back to her flat for coffee. After drinking their coffee they started to have sexual intercourse with Beth's true consent. However, just after Ahmed had penetrated Beth's vagina with his penis Beth said 'Stop, this is not right. I want us to just remain friends'. Ahmed put his hand over Beth's mouth and ignored her protests and continued to penetrate her vagina.

With regard to consent in s. 1, Sexual Offences Act 2003, which of the following statements is correct?

(a) The consent to the penetration operates only at the point when the penis enters the vagina. Therefore, Ahmed does not commit rape because Beth gives true consent at the time of penetration.

(b) Penetration is a continuing act from entry to withdrawal. Although Beth gives true consent on penetration, as soon as she withdraws her consent, Ahmed commits rape.

(c) It is an evidential presumption that Beth does not consent when she withdraws her consent.

(d) It is a conclusive presumption that Beth does not consent when she is induced to consent by a person known personally to her.

Answers to Self-assessment Test

1. (a) A reasonable person would consider the use of the vibrator as touching and sexual by its very nature.

2. (d) All the points to prove of rape are present. The mouth, vagina or anus can be penetrated in the offence of rape. The defendant does not have to ejaculate inside the victim for the offence to be committed, but is good evidence of penetration. The offence of assault by penetration does not include the mouth.

3. (c) The conclusive presumption in s. 76 refer to instances where the victim is deceived as to the nature or purpose of the relevant act or where they are induced to consent by impersonation of a partner. These are irrebuttable presumptions and cannot be rebutted. However, where violence has been used on the complainant there is an evidential presumption that Anne did not consent. The friend can rebut this presumption by adducing sufficient evidence to show she did consent and he reasonably believed it. Finally, in rebutting this evidential presumption, the friend would have to show he reasonably believed she consented and not just an honest belief on his part.

4. (b) Penetration is a continuing act and once consent is withdrawn any further penetration amounts to rape. The evidential presumption about consent does not include the withdrawal of consent. The conclusive presumption includes where the complainant is induced to consent by impersonating a person known personally to them.

Child Sex Offences

Objectives

With regard to child sex offences, at the end of this section, from a given set of circumstances, you will be able to:

1. Identify the points to prove for the offence of sexual activity with a child under 16 (s. 9, Sexual Offences Act 2003).

2. Identify the points to prove for the offence of causing or inciting a child under 16 to engage in sexual activity (s. 10, Sexual Offences Act 2003).

3. Identify the points to prove for the offence of sexual activity in the presence of a child under 16 (s. 11, Sexual Offences Act 2003).

4. Identify the points to prove for the offence of causing a child under 16 to watch a sexual act (s. 12, Sexual Offences Act 2003).

5. Identify the points to prove when offences under ss. 9 to 12 are committed by a person under 18 (s. 13, Sexual Offences Act 2003).

6. Identify the points to prove for offences when a person is in a position of trust (ss. 16 to 19, Sexual Offences Act 2003).

7. Identify the points to prove for the offence of arranging or facilitating the commission of child sex offences (s. 14, Sexual Offences Act 2003).

8. Identify the points to prove for the offence of meeting a child following sexual grooming (s. 15, Sexual Offences Act 2003).

9. Identify the points to prove for the offence of sexual activity with a child family member (ss. 25 and 26, Sexual Offences Act 2003).

10. Identify the points to prove for the offences involving indecent photographs of children (s. 1, Protection of Children Act 1978 and s. 160, Criminal Justice Act 1988).

11. Identify when the defences to offences involving indecent photographs of children apply.

Introduction

The Sexual Offences Act 2003 creates a number of offences under ss. 9 to 12, specifically to protect children from sexual offences committed by adults aged 18 or over. These replace previous child indecency legislation. Where the offender is under 18 the same offences (ss. 9 to 12) are committed under s. 13 but carry a lesser sentence. The Act also introduces some wide jurisdiction offences to deal with situations brought about by sex tourism and easier access to children by paedophiles through the internet.

Look at the first four offences and compare them by completing the table that follows.

Sexual activity with a child under 16 (s. 9, Sexual Offences Act 2003)

A person aged 18 or over (A) commits an offence if:

(a) he intentionally touches another person (B),

(b) the touching is sexual, and

(c) either:
    (i) B is under 16 and A does not reasonably believe that B is 16 or over, or
    (ii) B is under 13

Causing or inciting a child under 16 to engage in sexual activity (s. 10, Sexual Offences Act 2003)

A person aged 18 or over (A) commits an offence if:

(a) he intentionally causes or incites another person (B) to engage in an activity,

(b) the activity is sexual, and

(c) either:
    (i) B is under 16 and A does not reasonably believe that B is 16 or over, or
    (ii) B is under 13

Sexual activity in the presence of a child under 16 (s. 11, Sexual Offences Act 2003)

A person aged 18 or over (A) commits an offence if:

(a) he intentionally engages in an activity,

(b) the activity is sexual,

(c) for the purpose of obtaining sexual gratification, he engages in it:
    (i) when another person (B) is present or is in a place from which A can be observed, and
    (ii) knowing or believing that B is aware, or intending that B should be aware, that he is engaging in it, and

(d) either:
    (i) B is under 16 and A does not reasonably believe that B is 16 or over, or
    (ii) B is under 13

Causing a child under 16 to watch a sexual act (s. 12, Sexual Offences Act 2003)

A person aged 18 or over (A) commits an offence if:

(a) for the purpose of obtaining sexual gratification, he intentionally causes another person (B) to watch a third person engaging in an activity, or to look at an image of any person engaging in an activity,

(b) the activity is sexual, and

(c) either:
    (i) B is under 16 and A does not reasonably believe that B is 16 or over, or
    (ii) B is under 13

Section	Differences in points to prove	Common points to prove
9		1. 2. 3.
10		1. 2. 3.
11		1. 2. 3.
12		1. 2. 3.

Your table should have looked something like this:

Section	Sexual activity included in section	Common points to prove
9	A intentionally touches B.	1. The touching or activity has to be sexual. 2. Committed by a person (A) aged 18 or over, with or towards a child under 16 (B). 3. Where all the other ingredients of the offence are present and B is under 13, you only need to prove B's age. However, from the age of 13 to 16 you need to prove B's age and also that A did not reasonably believe that B is 16 or over.
10	A intentionally causes or incites B to engage in an activity.	As above.
11	A intentionally engages in an activity for purpose of obtaining sexual gratification. He engages in it when B is present or B is in a place where A can be observed and A knowing or believing that B is aware or intending that B should be aware that he is engaged in it.	As above.
12	For purpose of obtaining sexual gratification, intentionally causes B to watch a third person in an activity, or look at image of any person engaging in an activity.	As above.

You will see that the title of the middle column in this table has been changed to 'Sexual activity included in section'. This illustrates that the only difference between the sections is the type of sexual activity involved in the offence. The sections are intended to protect children under 16 from any form of sexual abuse. All the other points to prove are identical.

Position of Trust

The current legislation which protects children, from exploitation by people whom they should trust, has been consolidated into the Act.

Look at the following table of comparable offences committed by a person over 18 who is in a position of trust. Then identify the principle differences between ss. 16 to 19 and ss. 9 to 12 (above).

Section	Sexual activity included in section	Common points to prove
16	A intentionally touches B.	1. The touching or activity has to be sexual.
		2. A is in a position of trust in relation to B.
		3. It is assumed that A knows or could reasonably be expected to know of the circumstances by virtue of which he is in a position of trust in relation to B unless they prove otherwise.
		4. Committed by a person (A) aged 18 or over, with or towards a child (B) under 18.
		5. Where all the other ingredients of the offence are present and that B is under 13, you only need to prove B's age. However, from the age of 13 to 18 you need to prove B's age and also that A did not reasonably believe that B is 18 or over.
17	A intentionally causes or incites B to engage in an activity.	As above.
18	A intentionally engages in an activity for purpose of obtaining sexual gratification. He engages in it when B is present or B is in a place where A can be observed and A knowing or believing that B is aware or intending that B should be aware that he is engaged in it.	As above.
19	For purpose of obtaining sexual gratification, intentionally causes B to watch a third person in an activity, or look at image of any person engaging in an activity.	As above.

You will have noticed that because a position of trust may provide adults with greater opportunity to take sexual advantage of young people, the age of the child goes up to 18 years. Sections 16 to 19 would normally only be used in cases where the victim is 16 or 17.

You will also notice that it is to be taken that the defendant knew or could reasonably have been expected to know of the circumstances by virtue of which he was in such a position of trust unless sufficient evidence is adduced to raise an issue as to whether he knew or could reasonably have been expected to know of those circumstances.

This means that it will be accepted that a teacher at a school will know they are in a position of trust over all the pupils in the school unless the teacher can provide evidence to the contrary.

Who is a person in a position of trust in relation to another person? Section 21 details this as follows:

Where A looks after persons under 18 who:

- are detained in an institution by virtue of a court order or under an enactment, and B is so detained in that institution.
- are resident in a home or other place (s. 23(2) or 59(1) Children Act 1989) and B is resident, and is so provided with accommodation and maintenance or accommodation, in that place.
- are accommodated and cared for in; a hospital, an independent clinic, a care home, residential care home or private hospital, a community home, voluntary home or children's home, a residential family centre, and B is accommodated and cared for in that institution.
- are receiving education at an educational institution and B is receiving, and A is not receiving, education at that institution.

Other Child Sex Offences

These offences are designed to prevent paedophiles grooming children by gaining their confidence and then abusing them.

Look at the offences below.

Arranging or facilitating commission of child sex offences
(s. 14, Sexual Offences Act 2003)

A person commits an offence if:

(a) he intentionally arranges or facilitates something that he intends to do, intends another person to do, or believes that another person will do, in any part of the world, and

(b) doing it will involve the commission of an offence under any sections 9 to 13.

This could include people who knowingly arrange holidays abroad for paedophiles and procuring children for them with the specific aim of abuse. In this country it could amount to giving someone permission to use their house for sexual activity with a child.

Defence

There is a defence to this offence to avoid prosecuting professionals and adults, who aim to help and protect children from child pregnancy, physical harm, sexually transmitted diseases or while promoting emotional well-being by giving advice. This applies where it may be construed that the person may be arranging or facilitating an offence by providing contraceptives or giving advice on sexual matters, but in fact are intending to protect the child. Therefore, a teacher providing sex education to children would be covered by this defence.

Meeting a child following sexual grooming (s. 15, Sexual Offences Act 2003)

A person aged 18 or over (A) commits an offence if:

(a) having met or communicated with another person (B) on at least two earlier occasions, he:
    (i) intentionally meets B, or
    (ii) travels with the intention of meeting B in any part of the world,
(b) at the time, he intends to do anything to or in respect of B, during or after the meeting and in any part of the world, which if done will involve the commission by A of a relevant offence (Part 1 offences),
(c) B is under 16, and
(d) A does not reasonably believe that B is 16 or over.

Child abuse by strangers may have increased because there is greater accessibility to children through the anonymity of the Internet. There have been several high profile cases where paedophiles have groomed children on the Internet over a period of time before meeting them. This section would cater for these situations and applies to meetings anywhere in the world with the intention of abusing the child.

Child Sex Offences with Family Members

It is recognised that the vast majority of child sex offences occur in the family home or by people who know the child or their family. There are two offences of sexual activity with child family members to reflect the whole range of sexual abuse that can take place. Because family members may have a particular influence over other family members, these offences are very similar to the position of trust offences (see s. 16 above).

Look at the table below and identify the main difference between s. 16 and ss. 25 and 26.

Section	Sexual activity included in section	Common points to prove
25	A intentionally touches B.	1. The touching or activity has to be sexual. 2. A is related to B. 3. It is assumed that A knows or could reasonably be expected to know of his relation to B. 4. Committed by a person (A) aged 18 or over, with or towards a child (B) under 18. 5. Where all the other ingredients of the offence are present and that B is under 13, you only need to prove B's age. However, from the age of 13 to 18 you need to prove B's age and also that A did not reasonably believe that B is 18 or over.
26	A intentionally incites B to touch A or to allow himself to be touched by A.	As above.

You will notice that the offences are also committed against children under 18 and are identical to s. 16 except the person in position of trust is replaced by a family member over 18. The relationships above are drawn very widely to reflect today's varied family structure of blood relations and other family members.

Blood relatives can be:

- parents

- grandparents

- brothers, sisters

- half-brothers, half-sisters

- uncles, aunts

Other family members who are, or have been, living in the same household can be:

- foster parents

- foster siblings

- step-parents

- cousins

- step-brothers, step-sisters

- people who regularly care for a child and live in their household.

Other Offences with Family Members

Incest was always an offence between family members. It continues to be so under ss. 64 and 65 of the Act for blood relatives.

Section 64 makes it an offence for a person A, aged 16 or over, intentionally to penetrate sexually a relative B who is aged 18 or over if he knows or could reasonably have been expected to know that B is his relative.

Section 65 makes it an offence for a person A aged 16 or over to consent to being penetrated sexually by a relative B aged 18 or over if he knows or could reasonably have been expected to know that B is his relative.

Indecent Photographs of Children

You will be familiar with all the high profile publicity associated with offences of taking, distributing and possessing indecent photographs of children, particularly involving the Internet. Try the next activity to test your understanding of the legislation.

From your knowledge of current legislation answer 'true' or 'false' to the statements in the following quiz.

	Statement	true	false
1	It is an offence to take any indecent photograph or pseudo-photograph of a child under s. 1, Protection of Children Act 1978.		

	Statement	true	false
2	It is an offence for a person to have any indecent photograph or pseudo-photograph of a child in their possession under s. 2, Protection of Children Act 1978.		
3	A person will be a child for the purposes of the Act if it appears from the evidence as a whole that the child was, at the material time, under the age of 18.		
4	Pseudo-photographs of children will only be treated as a photograph where the impression conveyed is of an adult.		
5	There is a defence available for those taking indecent photographs of children, if done for a legitimate reason.		
6	There is a defence to the possession of indecent photographs of children that the photograph or pseudo-photograph was sent to the defendant without any prior request made by them and they did not keep it for an unreasonable time.		

Here are explanations together with the correct answer of true or false.

	Explanation	true	false
1	This was the first legislation to cover indecent photographs (s. 1(1)(a)). The section also includes: to distribute (s. 1(1)(b)), possess with a view to distribute (s. 1(1)(c)) or publish any advert to distribute (s. 1(1)(d)).	x	
2	It is an offence to possess indecent photographs of children, but is covered by s. 160(1), Criminal Justice Act 1988.		x
3	Both Acts now cover 16 and 17 year olds. This means there needs to be a conditional defence with regard to indecent photographs (not pseudo-photographs) where the child is over 16 and married or is a partner in an enduring family relationship. The conditions to the defence are that only the child or the defendant and the child are shown in the photograph and that the child consented (s. 1A and s. 160A).	x	
4	The correct interpretation of a pseudo-photograph: If the impression conveyed by a pseudo-photograph is that the person shown is a child or where the predominant impression is that the person is a child, that pseudo-photograph will be treated for these purposes as a photograph of a child, notwithstanding that some of the physical characteristics shown are those of an adult.		x

	Explanation	true	false
5	The defence to s. 1(1)(a) is much narrower than for legitimate reasons. It covers making photographs to prevent, detect or investigate crime and also covers the activities of the Security Service (s. 1B(1)). The defence for 'legitimate reasons' comes under s. 1(4) but not for taking indecent photographs. The defence only applies to s. 1(1)(b) and (c) which would cover exhibit officers at court. The defence is for the defendant to prove and also includes people who did not know they had indecent photographs in their possession.		x
6	Defences to possession under s. 160(1) are covered under s. 160(2) and are identical to the two defences under s. 1(4) above (explanation 5). However, there is a third defence to possession (as shown in statement 6 in the table above) under s. 160(2)(c).	x	

How did you get on? If you look at the legislation (s. 1, Protection of Children Act 1978 and s. 160, Criminal Justice Act 1988) in the **Investigator's Manual**, you will notice that the offences are fairly straight forward;

- taking (s. 1)
- distributing (s. 1)
- possessing to distribute (s. 1)
- publishing adverts to distribute (s. 1)
- possession (s. 160)

The defences are a little more complicated because they have been developed over time to ensure children are properly protected while other innocent people can go about their work or life unhindered (investigators, married couples, innocent recipients of unsolicited photographs).

Here is a schedule of the main offences together with the penalties and mode of trial. You will notice they are nearly all arrestable offences by virtue of having a penalty of 5 years' imprisonment.

Schedule of offences, their penalties and mode of trial

Offence	Maximum Penalty	Court
Non-consensual offences		
1 Rape	Life	Indictment
2 Assault by penetration	Life	Indictment
3 Sexual assault	10 years	Either way
4 Causing a person to engage in sexual activity without consent	Life, where penetration involved, otherwise 10 years	Where penetration involved, indictment only; otherwise either way

Offence	Maximum Penalty	Court
Rape and other offences against children under 13		
5 Rape of a child under 13	Life	Indictment
6 Assault of a child under 13 by penetration	Life	Indictment
7 Sexual assault of a child under 13	14 years	Either way
8 Causing or inciting a child under 13 to engage in sexual activity	Life, where penetration involved, otherwise 14 years	Where penetration involved, indictment only; otherwise either way
Child sex offences		
9 Sexual activity with a child	14 years	Where penetration involved, indictment only; otherwise either way
10 Causing or inciting a child to engage in sexual activity	14 years	Where penetration involved, indictment only; otherwise either way
11 Engaging in sexual activity in the presence of a child	10 years	Either way
12 Causing a child to watch a sexual act	10 years	Either way
13 Child sex offences committed by children or young persons	5 years	Either way
14 Arranging or facilitating the commission of a child sex offence	14 years	Either way
15 Meeting a child following sexual grooming etc	10 years	Either way
Abuse of position of trust		
16–19 Abuse of a position of trust offences	5 years	Either way
Familial child sex offences		
25 Sexual activity with a child family member	14 years where the defendant is aged 18 or over; otherwise 5 years	Where penetration involved, indictment only; otherwise either way
26 Inciting a child family member to engage in sexual activity		
Indecent photographs of children		
45 s. 1, Protection of Children Act 1978	10 years	Either way
s. 160, Criminal Justice Act 1988	5 years	Either way

Offence	Maximum Penalty	Court
Preparatory offences		
61 Administering a substance with intent	10 years	Either way
62 Committing an offence with intent to commit a sexual offence	Life where the offence is kidnapping or false imprisonment; otherwise 10 years	Indictment only where the offence is kidnapping or false imprisonment; otherwise 10 years
63 Trespass with intent to commit a sexual offence	10 years	Either way
Sex with an adult relative		
64 Sex with an adult relative: penetration	2 years	Either way
65 Sex with an adult relative: consenting to penetration	2 years	Either way
Other offences		
66 Exposure	2 years	Either way
67 Voyeurism	2 years	Either way
69 Intercourse with an animal	2 years	Either way
70 Sexual penetration of a corpse	2 years	Either way
71 Sexual activity in a public lavatory	6 months/level 5 fine	Summary

Self-assessment Test

1. Jim, aged 19, was baby-sitting for his neighbours. During the evening Claire, aged 12, came and sat next to him on the sofa to watch TV. When Claire fell asleep, Jim put his hand inside Claire's pyjamas and stroked the outside of her vagina.

   With regard to s. 9, Sexual Offences Act 2003, which of the following statements is correct?
   (a) There is an evidential presumption that Claire did not consent because she was asleep.
   (b) Jim is not guilty of sexual activity with a child under 16 if he reasonably believed Claire was 16.
   (c) Because Claire is 12 years old you do not have to prove Jim did not reasonably believe Claire was 16 or over.
   (d) Jim does not commit an offence because he is not yet 20 years old.

2. Carole, aged 42, knew that a schoolboy (aged 15), who lived opposite her house, regularly sat at his bedroom window looking into the street. One evening, when she saw him at his usual place, she ran upstairs into her bedroom opposite and turned on the light. She then intentionally started to undress provocatively until she was naked intending that the schoolboy should be aware of her undressing. She did this to get the thrill of exposing herself to the schoolboy.

   With regard to ss. 11 and 12, Sexual Offences Act 2003, which of the following statements is correct?
   (a) Carole is not guilty of sexual activity in the presence of a child because the child is not actually present in her bedroom.

    (b) Carole would not be guilty of sexual activity in the presence of a child if she reasonably believes the schoolboy is 16 or over.

    (c) Carole cannot be guilty of either offence because these offences can only be committed by a man.

    (d) Carole is guilty of causing a child to watch a sexual act.

3. Mr Jones, aged 25, is a science teacher at the local secondary school. Niki, aged 17, attends the school as a full-time student but has never been taught by Mr Jones. They both go on a school skiing trip. On the last night of the trip, after a moderate drinking session and with her true consent, Mr Jones inserts his penis into Niki's mouth.

    With regard to the Sexual Offences Act 2003, which of the following statements is correct?

    (a) Mr Jones commits no offence because Niki is over the age of consent and fully consents to the sexual activity.

    (b) Mr Jones is not in a position of trust with relation to Niki because he does not teach her.

    (c) Mr Jones is in a position of trust in relation to Niki but because she is over 16 he commits no offence.

    (d) Mr Jones is in a position of trust in relation to Niki and would have a defence to certain offences under the Act if he were acting to protect Niki.

4. Jose, aged 30, lives in London and gets a thrill by talking to children in chat-rooms on the Internet by pretending to be 16. He has no intention of meeting the children. One 14-year-old girl started to flirt with him during a series of separate conversations and she suggested that they should meet to have sex. Jose arranged for him and the girl to meet in Paris. Jose knew that Carl, aged 18 and who had a reputation for having underage sex with girls, would be interested in meeting the girl. Jose arranged for Carl to go to the meeting in his place. At the last minute the girl lost her courage and told her mum. Her mother contacted the police who arranged for the person meeting her daughter to be arrested when he arrived for the meeting.

    With regard to consent in ss. 14 and 15, Sexual Offences Act 2003, which of the following statements is correct?

    (a) No offences are disclosed in this country because Paris is outside our jurisdiction for the purposes of the Act.

    (b) Jose commits the offence of meeting the girl following sexual grooming, even though he has no intention of ever meeting her himself.

    (c) Jose commits the offence of arranging a child sex offence for Carl because he believes Carl will have sexual activity with the girl.

    (d) Carl commits the full offence of meeting the girl following sexual grooming even though Jose did the actual grooming.

5. With regard to offences under s. 1, Protection of Children Act 1978, and s. 160, Criminal Justice Act 1988, which of the following statements is *incorrect*?

    (a) If the impression conveyed by a pseudo-photograph is that the person shown is a child or where the predominant impression is that the person is a child, that pseudo-photograph will be treated for these purposes as a photograph of a child, notwithstanding that some of the physical characteristics shown are those of an adult.

    (b) It is a defence that there is a legitimate reason for distributing, showing, or possessing indecent photographs of children.

    (c) A person will be a child for the purposes of both Acts if it appears from the evidence as a whole that he/she was, at the material time, under the age of 16.

    (d) It is an offence to possess any indecent photograph or pseudo-photograph of a child.

Answers to Self-assessment Test

1. (c) Consent is not an issue in the offence under s. 9 with children. There is an evidential presumption with regards consent for adult victims under sexual assault by touching (s. 3). However, when the victim is not yet 13, under s. 3 consent is not an issue. In fact, there is a further offence of sexual assault by touching a child under 13 under s. 7, which increases the penalty from 10 to 14 years' imprisonment.

   The offence does not allow Jim to reasonably believe that a child under 13 is 16 or over.

   The age from which this offence can be committed is 18 or over.

2. (b) All the ingredients for sexual activity in the presence of a child under 16 are present. The offence can be committed by any person, man or woman (18 or over). The child does not have to be actually present with Carole, provided the child is in a place from which they can observe Carole. If Carole reasonably believed the schoolboy to be 16 or over she is not guilty of the offence even though the schoolboy is under 16.

   There is no third party present for the offence of causing a child to watch a sexual act.

3. (d) There is a defence to the offence of arranging or facilitating the commission of a child sex offence. These apply to teachers, doctors etc. who are acting to protect children from sexually infected diseases, child pregnancy etc. When Mr Jones is giving sex education lessons the defence could apply. Clearly, such a defence would not apply to Mr Jones in the circumstances of this question. Mr Jones can commit an offence under ss. 16 to 19 where he is in a position of trust as a teacher, even when Niki consents. Mr Jones does not need to directly teach Niki to be in a position of trust. He looks after people under 18 who are receiving education at the school and Niki is receiving, and Mr Jones is not receiving, education at that school. Offences under ss. 16 to 19 are specifically intended to protect 16 and 17 year olds.

4. (c) To commit the offence of arranging or facilitating Jose only needs to believe Carl will commit an offence under ss. 9 to 13. These two offences are wide ranging and can be committed here whilst involving other countries, provided there are similar sexual offences in the other country. Jose would need to intentionally meet or travel to meet the girl to do anything involving a relevant offence. Carl could commit the offence of attempting to meet a child following sexual grooming, but he would have needed to have met or communicated with the girl on at least two earlier occasions.

5. (c) The age has been increased from under 16 to under 18 years to give protection to 16 and 17 year olds.

PART FOUR

Evidence

Evidence

Objectives

With regard to evidence, at the end of this section, from a given set of circumstances, you will be able to:

1. Distinguish between a rebuttable presumption, an irrebuttable presumption and a presumption of fact.

2. Identify when documents are admissible by virtue of ss. 23 and 24, Criminal Justice Act 1988.

3. Identify what conditions must be satisfied when statements are served under s. 9, Criminal Justice Act 1967.

4. Identify when a fact can be formally admitted (s. 10, Criminal Justice Act 1967).

5. Identify the points to prove for intimidation and reprisals of witnesses, jurors etc. (s. 51, Criminal Justice and Public Order Act 1994).

6. Distinguish between intimidation and perverting the course of justice (common law).

7. Identify the points to prove perjury (s. 1, Perjury Act 1911).

8. Identify the points to prove for concealing arrestable offences (s. 5, Criminal Law Act 1967).

Introduction

The sources of criminal law are found partly in common law and partly in statute law.

Common law originated from the customs of the early communities, which were unified and developed in the Royal Courts during the three centuries following the Norman Conquest. Common law can be declared only an authority by the judges of the highest courts (Court of Appeal, High Court and House of Lords).

Legislation, in the form of Acts of Parliament, has always been regarded as supplementary to the common law. The majority of criminal law and procedure is now in statutory form. However, much of the principles of common law have over time been incorporated into legislation as appropriate.

Decided cases are decisions made by courts on the facts of a particular case and usually 'reported' in legal journals. The lower courts (magistrates' courts and Crown Courts) are bound by the decisions made by the Court of Appeal. The Court of Appeal is in turn bound by the decisions of the House of Lords, which is bound, to an extent, by decisions of the European Court of Justice.

The Human Rights Act 1998 introduces into English law the rights protected by the European Convention for the Protection of Human Rights and Fundamental Freedoms. The 1998 Act creates a statutory general requirement that all past and future legislation be read and given effect in a way that is compatible with the Convention.

General Principles

The judge determines questions of law.

The jury determines questions of fact.

So, the meaning of the words used in any Act of Parliament is a question of law, which the judge explains to the jury. The evidence that the jury hears contains the questions of fact, from which they decide innocence or guilt.

Now let us look at the English legal system. The words 'beyond all reasonable doubt' ring loudly when spoken in court. It sometimes appears an almost impossible burden. Quite clearly though, these words identify the legal system we work under.

There are two basic legal systems found in different parts of the world. Underline which one we use:

Inquisitorial

Accusatorial

The right answer is accusatorial. In other words, the state accuse a named person and put evidence before a court to prove guilt. This differs from an inquisitorial system. An inquisitorial system relies upon the questioning of all persons involved in an incident, to find out the truth and point out guilt. You could compare this with an inquest in this country.

Case Study 7—Albert Turner

Albert is separated from his wife. His friend Jim lives alone and for a short while allowed Albert to stay in his house, particularly while he was on bail for an offence of theft at the local off-licence. Albert stayed for about three months and then moved to a bedsit in the Bildston area of Sandford. After a further six months, Jim went to a locked cupboard where he kept his Sandford Building Society Savings Account Book. Jim couldn't find the book which had contained deposits of £3,000. Jim became worried. On further investigation, he couldn't find his driving licence either and Albert was the only person who knew where Jim hid the key to the cupboard. There was no sign of force to the cupboard. Jim went to his branch in Sandford. There, they checked the computer records and told Jim his money had all been withdrawn. He reported the matter to you. He tells you that he suspects Albert of stealing the book and withdrawing his money.

Consider the circumstances so far. Identify what type(s) of evidence Jim can give you at this stage to indicate whether Albert stole his savings book. Tick the appropriate answer(s) below.

(a) circumstantial ☐

(b) real ☐

(c) original ☐

(d) documentary ☐

Real evidence usually takes the form of a material object for inspection by the court. This evidence is to prove, either that the material object in question exists, or to enable the court to draw an inference from its own observation as to the object's value and physical condition. Such material objects are usually referred to as exhibits.

Original (primary) evidence is where a witness gives evidence to the court directly from the witness box. The evidence is presented to the court as evidence of the truth of what the witness states. Here the witness gives direct testimony about a fact of which they have personal or first-hand knowledge and therefore can be challenged on the truth of that fact in cross-examination.

Documentary evidence consists of documents produced for inspection by the court, either as items of real evidence or as hearsay or together with original evidence. Here the word 'document' includes maps, plans, graphs, drawings, photographs, discs, tapes, videotapes and films.

You should have concluded that Jim can only give circumstantial evidence. He has not seen the offence committed and it is only the events linking together that point to Albert being responsible for stealing the book. Look at the following definition of circumstantial evidence and see if the circumstances fit it so far.

Circumstantial evidence is evidence of relevant facts from which the facts in issue may be presumed with more or less certainty.

Some examples of circumstantial evidence include; facts that may supply a motive, facts that show planning to a subsequent action, a person's mental or physical capacity to do a particular act or evidence of opportunity.

This leads us to look at some basic rules.

Rules of Law

In our legal system, there are certain rules of law where the court accepts facts as proved when deciding whether the accused is guilty. These rules are called 'presumptions of law'. They are split into three types.

(a) irrebuttable presumptions;

(b) rebuttable presumptions;

(c) presumptions of fact.

An *irrebuttable presumption* (also called conclusive presumption) is where a court *must* accept the existence of certain basic facts even when evidence is brought to the contrary. In other words, in this instance a presumed fact can *never* be rebutted.

A *rebuttable presumption* is where a court *must* accept the existence of certain basic facts, unless evidence is given by the other party to prove that the presumed fact does not exist. In other words, in this instance a presumed fact can be rebutted.

A *presumption of fact* is where a court *may*, after evidence is given about certain facts, presume that another fact exists. In other words, a presumed fact may provide circumstantial evidence to infer that another fact is true.

Below are different presumptions. Put a tick against each presumption in the column you think appropriate.

Rule	Rebuttable	Irrebuttable	Presumption of fact
1 A child under 10 years cannot commit an offence.			
2 All persons in law are sane.			
3 A person is dead if unheard of for seven years.			
4 Albert Turner had the opportunity and the jury may presume that he stole the savings book (circumstantial evidence).			
5 When a person is in possession of recently stolen goods the court may infer that they are either the thief or receiver.			

The answers can be found in the following examples.

Irrebuttable presumptions. It is presumed that a child under 10 years cannot commit an offence.

Rebuttable presumptions. It is presumed that all persons are sane and a person is dead if unheard of for seven years.

Presumption of facts. It is presumed that when a person is in possession of recently stolen goods the court may infer that they are either the thief or receiver. The court may, after evidence is given about certain facts, presume (in the absence of sufficient evidence to the contrary), that another fact exists (circumstantial evidence).

Let us continue with the case of Albert.

You make enquiries through the Building Society Head Office. You are told by the head of security that the £3,000 has been systematically withdrawn in £50 cash amounts over the last six months, except for one large withdrawal of £1,500. This sum was withdrawn by way of a cheque to a third party. All the £50 sums were withdrawn at Newcastle and Gosforth branches of the building society alternately. He says he is unable to provide specific information and he was only told of the position by his branch staff.

Further enquiries take you to the building society branches themselves. There, you interview the cashiers. Three cashiers are able, using branch records, to tell you the dates and amount of withdrawals. One other cashier can also give you a description of the suspect and would be able to identify the man on an identification parade.

Which, if either, of the following statements is/are correct?

1. The three cashiers above can only give you circumstantial evidence of Albert's dealings in their branches.

2. The other cashier, identifies Albert, and can give you direct evidence of Albert's dealings in her branch.

   (i) 1 only   (iii) Both

   (ii) 2 only   (iv) Neither

Answer:...

The answer is (ii), only the second statement is correct. The cashier can give you direct evidence of both Albert's dealings and identification. The three other cashiers can, in fact, give direct evidence of his dealings, by making reference to their records of them.

When the cashiers cannot reasonably expected to remember the dealings with Albert, their evidence can be admitted as a business record.

Business records are admissible as evidence, by virtue of s. 24, Criminal Justice Act 1988, without the maker of the record having to attend court personally. This is allowable provided:

(a) the document was created or received by a person in the course of a trade, business or other occupation or as the holder of a paid or unpaid office; and

(b) the information contained in the document was supplied by a person (whether or not the maker of the statement) who had or may reasonably be supposed to have had personal knowledge of the matters dealt with.

The document will be admissible where the person who made the statement cannot reasonably be expected to have any recollection of the matters dealt with in the statement.

Similarly, where a witness cannot be found, their evidence can be admitted in evidence as follows.

*Documentary record*s. It is not always necessary for a witness to give evidence in person. By virtue of s. 23, Criminal Justice Act 1988, where a witness is unable to give original evidence in person, a statement made by that witness in a document will be admissible as evidence. This could apply where:

(a) the person is dead or unfit to attend as a witness; or

(b) the person is outside of the UK and it is not reasonably practicable to attend; or

(c) where the witness cannot be found.

This could also apply where:

(a) a witness makes a statement to an investigator; and

(b) the witness does not give evidence through fear.

Where such documents are admitted in evidence, the judge is required to warn the jury that the evidence has been admitted but that the defence has not had an opportunity to cross-examine the witness.

Let us go back to the case of Albert. Your enquiries reveal that the £1,500 cheque withdrawn from one of the branches was made payable to Value Cars (Sandford) Ltd. When you interview the manager of the company, the manager said that Albert Turner paid for a car by way of the cheque. The manager made a written statement.

Admissibility of Written Statements in Summary Proceedings

The evidence contained in a written statement can, in certain circumstances, be read out instead of the actual witness giving oral evidence from the witness box.

It can be used in summary trial under s. 9, Criminal Justice Act 1967. In such a case the witness need not be called to give evidence personally, but the statement of the witness is read out. Before these statements can be used as described, certain conditions have to be satisfied. Five of these conditions concern the actual layout and content of the statements. From your own knowledge, identify these by trying the next activity.

Study the statement over the page and see where you think it might fail to satisfy the conditions laid down under s. 9. List below the five conditions broken.

....

....

....

....

....

....

Westshire Constabulary

Statement of Witness

(C J ACT, 1967 ss 2, 9: M C Rules 1968 r58)

Statement of .... James OSBOURNE ....

Age of witness (Date of birth) .... Under 18 ....

Occupation of Witness .... Street Trader ....

Dated the .... 31st .... day of .... August ....

Signed ....

Signature witnessed by: ....

Further to my previous statement dated 20 August I was cleaning the room I let to Albert Turner when he stayed with me. While I was doing so, I lifted a loose floor board and found my driving licence. It is the same one I had locked in my cupboard with my savings account book. I have obtained a replacement licence and so I can produce it as evidence.

James Osbourne, being unable to read, I Detective Constable Jones read the above statement to him. When this was completed James Osbourne signed the declaration and statement.

Signed ....

Signature witnesed by: ....

You should have noticed the following faults in the statement:

1. It is not signed by the person who made it.

2. It contains no declaration as to perjury (Criminal Justice Act 1967).

3. It does not give the age of the witness, if under 18 years.

4. It does not contain the correct certificate regarding Jim's inability to read his statement. The certificate should read:

   (Name of Witness) being unable to read, I (Name of Officer) read the above declaration and statement to him/her and invited corrections, alterations and additions. When completed (Name of Witness) signed the declaration and statement.
   Officer's signature. . . . .

5. It refers to an exhibit (driving licence), which should be given a unique reference number using the initials of the witness together with a consecutive number and fully described therein.

All of these five mistakes cover the conditions required for the use of statements under s. 9, Criminal Justice Act 1967 and s. 102, Magistrates' Courts Act 1980. The conditions of service are:

1. The statement must be signed by the person who made it.

2. The statement must contain the declaration as to perjury (Criminal Justice Act 1967).

3. The age of the witness must be given if under 18 years.

4. The statement must contain the certificate, if it was read to him because he could not read.

5. Any document or exhibit referred to in the statement as an exhibit must be identified and fully described, with a label attached to that item; the label should have the description of the item as well as the signature of the witness and officer together with the unique reference number.

The defence are entitled to have statements etc., served on them at least seven days before summary trial. The prosecution are also entitled to have statements served on them at least seven days before summary trial.

The seven day period allows the defence or prosecution time to study evidence in the statement that is likely to be accepted by both parties. If, while reading the statements, either party feels that they would like to cross-examine a witness, that party can ask that the witness gives evidence in person.

If the service of a witness statement on the defence is carried out and no objection is received within seven days, it can be assumed that the witness is not required at the trial.

What if for any reason, the service of statements does not fall within the rules? All is not lost. If the prosecution and defence agree, statements can still be read out instead of oral evidence being given. However, the court can require a witness to be called rather than allowing a statement to be read.

Now look at the rest of the conditions of service:

6. A copy of the statement is served before the trial on the defence (or prosecution if defence witness).

7. Service on the defence (to include copies of any documents referred to in the statement) can be done by personal service or by recorded delivery or registered post to (a) the solicitor or (b) the defendant's usual or last known place of abode.

8. No objection to the s. 9 procedure has been received within seven days of such service.

9. If service is impracticable, the s. 9 procedure can still apply, if both prosecution and defence agree.

10. The statement is read aloud by the party producing it (can be read in précis form).

Formal Admissions (s. 10, Criminal Justice Act 1967)

In all prosecutions, when the defence accepts a fact, there is a process by which this can be formally admitted under s. 10, Criminal Justice Act 1967. In such formal admissions the fact ceases to be an issue. Where the accused enters a not guilty plea at a plea directions hearing, both the prosecution and defence are expected to inform the court of the facts which are to be admitted.

Intimidation of Witnesses, Jurors and Others, and Reprisals against Them

Albert is eventually charged with theft and deception. While awaiting trial, he realises that the manager of Value Cars (Sandford) Ltd has made a witness statement and is due to give evidence at the trial. In an effort to weaken the case against him, Albert telephones the manager's partner at home and says, 'Unless your partner goes missing at the time of the trial I'll make sure you don't walk again.'

What offence has Albert committed?

....

....

....

....

You may have considered the common law offence of attempting to pervert the course of justice. However, s. 51(1), Criminal Justice and Public Order Act 1994, was brought in specifically to deal with an increasing trend of intimidation.

Section 51(1), Criminal Justice and Public Order Act 1994

A person commits an offence if—

(a) he does an act which intimidates, and is intended to intimidate another person ('the victim');

(b) he does the act knowing or believing that the victim is assisting in the investigation of an offence, or is a witness or a potential witness or a juror or potential juror in proceedings for an offence, and

(c) he does it intending thereby to cause the investigation or the course of justice to be obstructed, perverted or interfered with.

Once the fact that Albert threatened the manger's partner has been proven, it is necessary to prove that he knew or believed that the manager was assisting in the proceedings. This is an offence of 'specific intent'. Interestingly there are two elements of the offence where specific intent needs to be proved. So the prosecution need to prove that Albert made the threats intending to intimidate the manager. Then it is necessary to prove that the intimidation was intended to obstruct, pervert or interfere with the course of justice.

Any person can be intimidated and the intimidation does not need to be in the presence of the victim. The intimidation can be physical, emotional or financial.

Section 51(2), Criminal Justice and Public Order Act 1994, provides a similar offence for when reprisals are taken out on people after an investigation or prosecution. Here it is necessary to prove that the harm caused was intended to be caused or any threat of harm was intended to cause the victim to fear harm because the accused knew or believed the victim assisted in an investigation or proceedings. The harm done or threatened can be financial or physical.

Section 51(2), Criminal Justice and Public Order Act 1994

A person commits an offence if—

(a) he does an act which harms, and is intended to harm, another person or, intending to cause another person to fear harm, he threatens to do an act which would harm that other person,

(b) he does or threatens to do the act knowing or believing that the person harmed or threatened to be harmed ('the victim'), or some other person, has assisted in an investigation into an offence or has given evidence or particular evidence in proceedings for an offence, or has acted as a juror or concurred in a particular verdict in proceedings for an offence, and

(c) he does or threatens to do it because of that knowledge or belief.

These specific intimidation offences were introduced to ensure that witnesses feel confident that they can give evidence or jurors can deliver verdicts without fear of intimidation or later reprisals. There has always been a common law offence of perverting the course of justice.

Perverting the Course of Justice

Common Law

It is an offence at common law to do any act tending and intending to pervert the course of public justice.

This is a serious offence that carries a penalty of life imprisonment. This offence includes acts where evidence is deliberately destroyed, concealed or falsified as well as cases where witnesses or jurors are intimidated. Examples include making a false allegation of crime, pretending to be someone else when being dealt with for an offence, setting fire to evidence in possession of the police, etc.

A related offence that interferes with the course of justice is perjury.

Perjury

Section 1(1), Perjury Act 1911

If any person lawfully sworn as a witness or as an interpreter in a judicial proceeding wilfully makes a statement material in that proceeding, which he knows to be false or does not believe to be true is guilty of perjury . . .

The penalty for perjury is seven years' imprisonment.

Clearly, to commit the offence someone needs to be lawfully sworn, but a sworn statement (affidavit) read out in court would also be lawfully sworn. There is a lesser offence (two years' imprisonment) when a false statement is tendered to the court under the Criminal Justice Act 1967.

Concealing Arrestable Offences

This offence (s. 5, Criminal Law Act 1967) is committed when someone strikes a deal with an offender to withhold information from you, the investigator, which might be of material assistance in securing a conviction against the offender. There are five points you would need to prove:

1. That an arrestable offence has been committed.

2. That the accused knew or believed the offence, or some other arrestable offence, had been committed.

3. That the accused knew or believed that they had information which might be of material assistance in securing the prosecution or conviction of an offender for the offence.

4. That the accused accepted, or agreed to accept, any consideration (other than the making good of loss or injury or making reasonable compensation).

5. That the accused accepted, or agreed to accept, that consideration in exchange for not disclosing the information.

Once you have proved these points, you will need to liaise with the CPS in order to get the consent of the Director of Public Prosecutions before proceedings are instituted.

Summary

This section was intended to give you a short introduction to the subject of evidence.

It should have brought back to mind some of the aspects of evidence you have already learned, but possibly forgotten.

The latter section should have introduced you to offences that address interfering with the course of justice.

Self-assessment Test

Having completed this section, test yourself against the objectives at the beginning of the section. You will find the answers below.

1. Which of the following is an irrebuttable presumption of law?
    (a) All persons in law are sane.
    (b) A person is dead if unheard of for seven years.
    (c) A child under 10 cannot commit an offence.
    (d) When a person is in possession of recently stolen goods the court may infer that they are either the thief or receiver.

2. With regard to s. 23, Criminal Justice Act 1988, which of the following could *not* be admitted in evidence as a documentary record?
    (a) A statement made by a person who is unfit to attend court as a witness.
    (b) A person who is outside the UK and it is not reasonably practicable for them to attend.
    (c) A statement made by a witness who cannot be found.
    (d) A person who makes a statement and is unwilling to give evidence.

3. With regard to s. 9, Criminal Justice Act 1967, which of the following is *incorrect*?
    (a) A statement must be signed by the person who made it and contain the declaration as to perjury.
    (b) Any document referred to in the statement as an exhibit must be readily available at court.
    (c) Where the person making a statement cannot read, the statement must contain a certificate to that effect and what has been done when making the statement.
    (d) The statement must contain the age of the witness if under 18 years.

4. With regard to s. 51, Criminal Justice and Public Order Act 1994, which of the following is *not* an offence of intimidation.
    (a) Simon was been charged with actual bodily harm. He threw a brick through the window of Sarah's house intending to prevent her giving evidence against him at his trial.
    (b) Keith committed a robbery of a jeweller's shop. Before the alarm could be raised and on his way out of the shop, he pointed a gun at the security guard and said 'If you give evidence I'll be back'.

(c) Liam had become a suspect for a robbery he had committed 15 months earlier. He heard that the National Crime Squad was now investigating the offence. He sent a bouquet of flowers to the victim of the robbery intending to stop her from giving evidence against him. On a greetings card with the flowers he wrote, 'I hope you have plenty of photographs of your face, because you will need them if you give evidence.'

(d) Outside the courtroom John went up to Freddie who had just given evidence against John at his trial for indecent assault. John said, 'I'm going to get you for what you've done'.

Answers to self-assessment Test

1. Answer (c). Where the fact that a person is under ten years old is accepted by a court, the fact that they cannot commit an offence is irrebuttable.

2. Answer (d). The fourth category is where the witness is too frightened to give evidence. Special measures are now available to enable vulnerable witnesses to give evidence more effectively.

3. Answer (b). Any document or exhibit referred to in the statement as an exhibit must be identified and fully described etc.

4. Answer (b). Although Keith intimidates the security guard, neither an investigation nor proceedings have yet started.

Right to silence

Objectives

With regard to right of silence (ss. 34–37, Criminal Justice and Public Order Act 1994), at the end of this section, from a given set of circumstances, you will be able to:

1. Identify when an inference can be drawn from the silence of a suspect under (s. 34).

2. Identify what constitutes a significant statement or silence.

3. Decide whether the criteria have been satisfied for an inference to be drawn from the failure or refusal to account for objects, substances or marks (s. 36).

4. Decide whether the criteria have been satisfied for an inference to be drawn from the failure or refusal to account for presence at a particular place (s. 37).

5. Identify the three points that need to be included in a warning to the suspect (ss. 36 and 37).

Introduction

An area of law that is closely associated with evidence is the suspect's right of silence and the effect in law of a failure to answer questions or a failure to give evidence in a person's own defence. Under certain circumstances a magistrate or jury can draw their own conclusion why a person has failed to account for a fact.

You may feel that access to Code C of the PACE Codes of Practice will be advantageous to you completing this section successfully. Code C is set out in the **Investigator's Manual**.

The Criminal Justice and Public Order Act 1994 permits a court to draw such inferences as are appropriate in four sets of circumstances:

(a) where a suspect fails, when questioned under caution or charged, to mention facts later relied on as part of his or her defence, when it is reasonable to expect him or her to have mentioned them; (s. 34)

(b) where an accused fails without due cause to give evidence or answer questions at trial; (s. 35)

(c) where an arrested person fails or refuses to account for possession of objects, substances or marks when requested to do so; (s. 36)

(d) where an arrested person fails or refuses to account for presence at a particular place, when requested to do so (s. 37).

It is important that you satisfy the legal criteria to ensure that where there is a silence, as defined in the Act, it can be put before the jury. The following illustrates such considerations.

Relevant Facts not Mentioned (Section 34)

Write down the relevant points which need to be satisfied before an inference can be drawn at court in relation to relevant facts when spoken to by the investigator.

(a) ....

(b) ....

(c) ....

or

(d) ....

Your answer should include something similar to the following:

(a) when questioned must be under caution;

(b) it must be before charge;

(c) before charge the person failed to mention any fact later relied on in their defence; or

(d) on being charged or told that they will be prosecuted, failed to mention any such fact.

A 'fact not mentioned' is one that the accused could reasonably have been expected to mention when questioned or charged. 'Fact' is defined in s. 34(1) should you require more information.

These facts could be defences, so consider some in your own mind. Here are a few examples: alibi, accident and self-defence (in cases of assault). The list is endless, but relevant to each situation.

In order to make the accused aware of possible inferences that could be drawn, the caution is used.

Caution

The full caution should read:

> You do not have to say anything. But it may harm your defence if **you do not mention when questioned** something which you later rely on in court. Anything you do say may be given in evidence.

The 'caution' is obviously an essential part of an interview. Without it, an interview is almost bound to be inadmissible. It is your responsibility (not the solicitor's) to ensure that the suspect understands the caution. If you are in any doubt whether they do, you should explain it to them in your own words.

For instance, the caution can be explained to the suspect in three parts:

(a) The right of silence has not been abolished and you always have the right to say nothing.

(b) If the case goes to court and your defence makes use of something which you do not mention now, and in the circumstances, the court thinks you should have mentioned it now, then the court can draw its own conclusion of how much notice to take of the new information.

(c) If you do answer questions, the words spoken will be recorded and read out or played in court.

The caution at the time a person is charged or told that they are going to be prosecuted differs slightly from the caution shown above. Can you write down reason(s) for this?

....

....

....

....

This caution reads:

> You do not have to say anything. But it may harm your defence **if you do not mention now** something which you later rely on in court. Anything you do say may be given in evidence.

The difference occurs because the person is being charged and no questions should be asked at this stage. There are circumstances when you can interview the charged person further about the offence and they are:

- to prevent or minimise harm or loss to some other person, or the public

- to clear up any ambiguity in a previous answer or statement

- it is in the interest of justice for the detainee to have put to them, and have an opportunity to comment on, information concerning the offence which has come to light since they were charged (PACE Code of Practice C 16.5).

Significant Statements and Silences

At the beginning of an interview under caution at a police station, any significant statement or silence which occurred prior to arrival at the station shall be put to the suspect and he or she should be asked whether they confirm or deny it and if they wish to add anything to it.

Think what a significant statement might be.

....

....

....

....

....

....

You might have recollections of situations where things have been said under caution while you have been at the scene of an incident.

PACE Code C, para. 11.2A, defines a significant statement or silence, which includes one that appears capable of being used in evidence against the suspect. In particular:

(a) A direct admission of guilt.

(b) A failure to answer a question at all, after caution.

(c) A failure to answer satisfactorily a question put after caution.

(d) A refusal to answer a question at all after caution.

(e) A refusal to answer a question satisfactorily after caution.

Do not forget that the most important safeguard you should make in relation to these significant statements or silences is to ensure their integrity before bringing them up at a later interview.

You do this by making a record of the questions and answers and giving the suspect an opportunity to sign the record at the earliest opportunity, with the words: 'I agree that this is a correct record of what was said' (PACE Code C, Note 11D).

This approach will also apply to 'relevant' comments which do not fall within the significant statement definition and may even be made before caution. You will probably identify them as unsolicited comments but the important factor is that they are relevant to the offence. So if any of these type of comments occur outside the interview, deal with them in the same way as the significant ones defined above.

To summarise, s. 34 creates a situation where a court may draw an inference in determining a person's guilt if an *accused raises a defence at court* which relies on a fact that has not been mentioned before, provided he or she has been given the opportunity to mention that fact when questioned under caution; and it is reasonable to expect the suspect to have been able to have mentioned that fact at the time.

Failure or Refusal to Account for Objects, Substances or Marks (Section 36)

Write down the criteria which need to be satisfied before an inference can be drawn at court, of an accused's failure to account for objects, substances or marks.

....

....

....

....

....

....

Your answer should include:

(a) the person must have been arrested by a constable and there is:

    (i)   on the suspect's person, or

    (ii) in or on their clothing (which includes footwear), or

    (iii) otherwise in their possession, or

    (iv) in any place where they are at the time of their arrest,

    any object, substance or mark or there is any mark on any such object; *and*

(b) the arresting constable, or any other investigator, reasonably believes that the presence of the object, substance, or mark may be attributable to the participation of the person in the commission of an offence specified by the investigator; *and*

(c) the investigator informs the arrested person, of that belief and requests him or her to account for the presence of the object, substance, or mark; *and*

(d) that person fails or refuses to do so.

These then are the points which need to exist and be satisfied before a court can draw any later inference, so let us try and illustrate the procedure which needs to be adhered to make sure this section can apply.

Picture the following incident. You are called to the scene of a robbery where the victim and a witness relate to you the details and give a full description of the suspect which includes the facts that he will have a torn jacket sleeve where the victim made a grab for him and he is carrying a penknife. You commence a search and a short distance away in a shop doorway you find the suspect, out of breath, and who on seeing you throws a penknife into the corner. You caution him and tell him he is under arrest for robbery and notice then he has a torn jacket sleeve.

What questions can you ask him at the scene now he is under arrest?

(a) any questions at all because he is under caution;

(b) none because once an arrest is made, he should be interviewed only at a police station (with three exceptions);

(c) only questions concerning the penknife in the corner of the doorway;

(d) only questions concerning the torn sleeve.

The answer is (b).

PACE Code C, para. 11.1, states that where a decision to arrest is made, a person must not be interviewed other than at a police station unless the consequent delay would be likely:

(a) to lead to interference or harm to evidence or persons;

(b) to lead to the alerting of suspects not yet arrested;

(c) to hinder the recovery of property obtained from the relevant offence.

You may think that asking one question at the scene is not an interview but that depends on the definition of an interview.

So, how would you define an interview?

....

....

....

....

PACE Code C, para. 11.1A

An interview is the questioning of a person regarding his involvement or suspected involvement in a criminal offence or offences which, by virtue of paragraph 10.1 of Code C, is required to be carried out under caution . . .

In the example above, it would be fair to say that the question regarding the penknife or the torn sleeve would go some way towards indicating the suspect's involvement in the offence of robbery.

As you can see from this situation, s. 36 will apply in the main to the interview scenario at the police station and not immediately upon arrest.

Before moving on to the requirements which you need to fulfil before any later inferences can be made at court, we should look at s. 37 of the same Act. This is similar in nature to s. 36 but deals with different facts.

Failure or Refusal to Account for Presence at a Particular Place (Section 37)

This section is very similar in nature to s. 36, so bearing in mind the title of the section, write down the criteria which you think need to be satisfied before an inference can be made at court.

(a)...

....

(b)...

....

(c)...

....

(d)...

....

Your answer should include the following. Section 37 permits an inference to be made where:

(a) a person arrested by a constable was found by him or her at a place at or about the time of the offence for which the person was arrested is alleged to have been committed;

(b) that or another investigator reasonably believes that the presence of the person at that place and at that time may be attributable to his or her participation in the commission of the offence; *and*

(c) the constable or investigator informs the arrested person of that belief and requests him or her to account for that presence; *and*

(d) the person fails or refuses to do so.

As you no doubt can see, the wording is very similar to the previous s. 36 (particularly that the person is under arrest) but s. 37 relates to a person giving an explanation for their presence at a certain place at a certain time which more than likely will be the scene of an offence or arrest. (A place is described in the widest possible terms and includes buildings and parts of buildings, vehicles, vessels and aircraft etc.)

Now let us take our example of the robbery and arrest scenario forward to the interview at the police station. What are the facts which you would want to bring up at the interview in relation to objects, marks, presence etc.?

(a)...

....

(b)...

....

(c)...

....

From the circumstances, the following would be facts for which you would require an explanation in the interview.

(a) the penknife being thrown in the doorway (object in the place of suspect's arrest—s. 36(1)(a)(iv));

(b) the tear in his jacket sleeve (mark on his clothing—s. 36(1)(a)(ii));

(c) his presence in the doorway at the time of arrest close to the scene of the robbery (s. 37(1)(a)).

Before any inferences can be drawn from replies or silences of an accused in answer to questions about these facts, what do you as an interviewer have to do?

(Remind the suspect they are still under caution, entitled to free legal advice and the need to put any previous significant statements or silences to them.)

PACE Code C, para. 10.5B has the answer. Put briefly, you should at the start of the interview ensure that you remind the suspect of what offence you are investigating, that a record is being made of the interview and it may be given in evidence, if it reaches trial.

The next three points have to be put to the suspect at the 'challenge' stage when you know you are going to be asking about the relevant facts. They should be put in ordinary language so that the suspect understands what you are saying and they are:

(a) what fact it is you are asking the suspect to account for, e.g. the tear, the penknife or presence;

(b) that you believe the fact may be due to the suspect's taking part in the commission of the offence in question; and

(c) a court may draw a proper inference if the suspect fails or refuses to account for the fact about which he is being questioned.

This shows that the onus will be on you, as interviewer, to ensure that the requirements are fulfilled regarding the warnings. Remember they have to be in ordinary language.

Do not let this interrupt the structure of your interview. The need to plan is essential. If you get it wrong and the suspect does not understand, the consequences are likely to be that an inference will not be allowed in court.

Restriction on Drawing Adverse Inferences

Since the case of *Murray* v *United Kingdom* (1996), there is a restriction placed on courts from drawing an adverse inference on a suspect's silence. This means that when a suspect has their access to legal advice delayed, the court cannot draw an adverse inference from their silence (Code of Practice Annex C).

This restriction applies to:

(a) any detainee at a police station who before being interviewed or being charged or informed they may be prosecuted has:

  (i) asked for legal advice;

  (ii) not been allowed an opportunity to consult a solicitor, including the duty solicitor, as in this code; and

  (iii) not changed their mind about wanting legal advice

When this restriction applies there could be confusion in the mind of the suspect because the original arrest caution warns the suspect about remaining silent. The situation changes as legal advice is delayed and now the suspect's silence would not be subject to adverse inferences at court.

This situation is described in Code of Practice C Annex C Note C2. The suspect has previously been cautioned that it may harm their defence if they do not mention, when questioned, something they may later rely on in court. However, now that legal advice has been delayed, this has changed and no adverse inference can be drawn. Therefore, you will need to explain this to them and caution them with the 'old' caution. The reverse is the case when the restriction ceases to apply (legal advice allowed or suspect decides they do not want legal advice), then you will have to explain the further change in position and re-caution with the usual caution.

The following is suggested as a framework to help explain changes in the position on drawing adverse inferences when there is a restriction in place:

(a) Restriction begins to apply:

'the caution you were previously given no longer applies. This is because after caution:

(i) you asked to speak to a solicitor but have not yet been allowed an opportunity to speak to a solicitor,' or

(ii) you have been charged with/informed you may be prosecuted.

'This means that from now on, adverse inferences cannot be drawn at court and your defence will not be harmed just because you choose to say nothing. Please listen carefully to the caution I am about to give you because it will apply from now on. "You do not have to say anything unless you wish to do so but what you say may be given in evidence". You will see that it does not say anything about your defence being harmed.'

(b) Restriction ceases to apply or at the time the person is charged or informed they may be prosecuted.

'The caution you were previously given no longer applies. This is because after the caution you have been allowed an opportunity to speak to a solicitor/decided you don't now want a solicitor. Please listen carefully to the caution I am about to give you because it will apply from now on. It explains how your defence at court may be affected if you choose to say nothing' (the usual caution should then be applied).

Failure to Give Evidence at Trial (Section 35)

In your day-to-day investigation of crime, s. 35 is not likely to be of any relevance to you. However, it may be worth knowing that this section does create another opportunity for a court to draw inferences from an accused's silence, or refusal to testify in his or her own defence, or even a refusal to be sworn.

Summary

This section is intended to give you an insight into the legislation which enables courts to make inferences from an accused's silence provided investigators comply with certain conditions.

This area of law and procedure does not remove the suspect's basic right of silence. Some may see it as coercive (particularly the defendant), so conformity with the rules and the necessary warnings is vital so investigators can be seen to be fair and totally impartial.

Self-assessment Test

Having completed this section, test yourself against the objectives outlined at the beginning of the section. You will find the answers below.

1. With regard to the caution (s. 34, Criminal Justice and Public Order Act 1994), the caution can be divided into three parts. Which of the following part of the caution is *not* a correct explanation of that part?
   (a) The right of silence has not been abolished and you always have the right to say nothing.
   (b) If the case goes to court and your defence makes use of something which you do not mention now, and in the circumstances, the court thinks you should have mentioned it now, the court can draw its own conclusion and how much notice to take of the new information.
   (c) If the case goes to court and your defence makes use of something which you mention now but you change at court, and in the circumstances, the court thinks you should have mentioned it now, the court can draw its own conclusion and how much notice to take of the new information.
   (d) If you do answer questions, the words spoken will be recorded and read out or played in court.

2. With regard to PACE Code C, para. 11.2A, which of the following would *not* be considered a significant statement or silence?
   (a) Mike was caught committing a burglary in a warehouse by PC Grant. Before being arrested or cautioned Grant said, 'Alright, alright, alright I know I shouldn't be in here'.
   (b) Jo was stopped by PC Brooks on suspicion of stealing a gold ring. Before being arrested and cautioned PC Brooks said, 'Where have you just come from?'. Jo remained silent.
   (c) Nasser was arrested for grievous bodily harm by PC Kipp and said on being cautioned 'He got what he deserved'.
   (d) Jack was arrested at his home address by PC Bennett for handling stolen credit cards. After being cautioned he was asked 'Where are the cards hidden?' Jack remained silent.

3. With regard to s. 36, Criminal Justice and Public Order Act 1994, in which of the following circumstances would a special warning *not* be appropriate?
   (a) Steve was arrested at his home address by PC Thomas for a recent burglary. During the subsequent search of his home a video recorder stolen during the burglary was found in his living room. When interviewed at the police station Steve was asked to account for his possession of the video recorder. Steve made no reply.
   (b) Ahmed was arrested at Perrings garage, where he works as a mechanic. He was arrested by DC Phillips for handling a stolen TV set and taken back to his home address which was searched under s. 18, Police and Criminal Evidence Act 1984. The stolen TV was found in his garden shed. When asked during interview to account for the TV, Ahmed made no reply.
   (c) Information was received that Tony was responsible for a burglary and that he had sold the stolen property to 'Old Stuff' second-hand shop. PC Fuller arrested Tony for the burglary. A video recorder stolen from the burglary was then recovered from 'Old Stuff'. When interviewed at the police station and asked to account for the video recorder, he made no reply.
   (d) Graham was arrested by PC Shah at his home address for a burglary where a TV set was stolen. The stolen TV set was recovered from a cupboard under the stairs. At the police station during interview when asked to account for the TV set Graham said, 'A green

man with two heads and seven arms came through the window and put the TV in the cupboard and left.'

4. With regard to s. 37, Criminal Justice and Public Order Act 1994, in which of the following circumstances would a special warning *not* be appropriate?

    (a) During a burglary at a factory three weeks ago the alarm went off. PC Evans attended the scene and recognised Philip at the point of entry to burglary. Philip made good his escape and was circulated as wanted. Today Philip was seen by PC Grant and arrested. During interview by DC Brent at the police station the suspect was asked to account for his presence near the scene. Philip makes no reply.

    (b) During a burglary at a house the intruder was disturbed. PC Chang attended the scene and arrested Ben in the garden of the house. During interview by PC Chang at the police station Ben was asked to account for his presence at the scene of the burglary. Ben made no reply.

    (c) During a burglary at a factory on the outskirts of a town at 3 am the alarm went off. PC Strong attended the scene and arrested Wane about 50 metres from the building. During interview by PC Strong at the police station Wane was asked to account for his presence near the scene when arrested. Wane made no reply.

    (d) During a burglary at a factory three weeks ago the alarm went off. PC Gent attended the scene and recognised Lyle at the point of entry to the burglary. Lyle made good his escape. Today, PC Gent saw Lyle and arrested him. During interview by DC Frost at the police station, Lyle was asked to account for his presence near the scene. Lyle made no reply.

5. With regard to PACE Code C, para. 10.5B, which of the following is *not* one of the points that have to be stated when giving a special warning to a suspect during an interview.

    (a) That a court may draw a proper inference if he changes his account for the fact about which he is being questioned.

    (b) That a court may draw a proper inference if he fails or refuses to account for the fact about which he is being questioned.

    (c) What fact it is you are asking the suspect to account for.

    (d) That you believe the fact may be due to the suspect taking part in the commission of the offence in question.

6. With regard to the restriction on adverse inferences, which of the following action should be taken when you interview a suspect who has asked for legal advice but legal advice has been delayed?

    (a) You only have to tell the suspect that no adverse inference can be drawn from any of his answers.

    (b) You should tell the suspect that as soon as the delay in access to legal advice ceases the restrictions on adverse inferences will also cease.

    (c) You should explain to the suspect that the previous caution no longer applies because their access to legal advice has been delayed. This means that from now on no adverse inference can be drawn by the court and their defence will not be harmed if they choose to say nothing. Then caution the suspect as follows 'You do not have to say anything unless you wish to do so but what you say may be given in evidence'.

    (d) You should explain to the suspect that the previous caution applies even though their access to legal advice has been delayed. This means that adverse inference can be drawn by the court and his defence will be harmed if he chooses to say nothing. Then further caution the suspect as normal.

Answers to Self-assessment Test

1. Answer (c). The caution does not cover cases where a suspect changes his or her account at court.

2. Answer (b). A significant silence is a silence to a question after caution only.

3. Answer (c). When arrested, Tony was no longer in actual or constructive possession of the video recorder.

4. Answer (a). The officer finding Philip must be the same officer that arrests him for a special warning to apply.

5. Answer (a). Special warnings do not cover cases where a suspect changes his or her account about a fact at court.

6. Answer (c). When legal advice is delayed, a restriction applies on drawing adverse inferences from a suspect's silence. Therefore, a further caution needs to be used to avoid confusion in the mind of the suspect.

Bail

Objectives

With regard to bail, at the end of this section, from a given set of circumstances, you will be able to:

1. Identify the general right to bail.
2. Identify when the general right to bail does not apply.
3. Distinguish between the exceptions to bail.
4. Identify the types of conditions that can be applied to bail.
5. Identify the reasons for applying conditions to bail.
6. Identify the power of arrest under the Bail Act 1976.

Introduction

Imagine you had to write a dictionary definition of bail. What would you write? Spend a couple of minutes thinking about this, then write your definition in the space below:

....

....

....

....

Bail is a recognizance or bond taken by a duly authorised person to ensure the appearance of an accused person at an appointed place and time to answer the charge made against that person.

Does your definition consider different types of offences?

Right to Bail

Look at the list of offenders below. Indicate by a tick, which ones you think would be entitled to bail.

Bill the shoplifter ☐   Bert the rapist ☐

Betty the arsonist ☐   Ben the murderer ☐

Not all the above would be entitled to bail under s. 1, Bail Act 1976.

Section 25, Criminal Justice and Public Order Act 1994 states that where a person is charged with:

(a) murder, manslaughter, rape, *or*

(b) an attempt to murder or rape, *or*

(c) an offence under the Sexual Offences Act 2003, section 2 (assault by penetration), 4 (causing a person to engage in sexual activity without consent), 5 (rape of a child under 13), 6 (assault of a child under 13 by penetration), 8 (causing or inciting a child under 13 to engage in sexual activity), 30 (sexual activity with a person with a mental disorder impeding choice), or 31 (causing or inciting a person, with a mental disorder impeding choice, to engage in sexual activity), or an attempt to commit any of these offences, *and*

(d) has previously been convicted in the UK of one of those offences, *or*

(e) culpable homicide, *and*

(f) in the case of previous conviction for manslaughter or culpable homicide, he or she was sentenced to imprisonment,

he or she will be granted bail in exceptional circumstances only.

So if Bert or Ben had been previously convicted of any of the above offences, they will not be automatically entitled to bail.

General Right to Bail

Section 1(6), Bail Act 1976

Bail in criminal proceedings shall be granted (and in particular shall be granted unconditionally or conditionally), in accordance with this Act.

What this section is saying is that any person (apart from those previously mentioned in s. 25, Criminal Justice and Public Order Act 1994) involved in criminal proceedings is entitled to bail within the rules of the 1976 Act. Street bail is now available to enable you to manage investigations more effectively, by giving you the power to grant an arrested person immediate bail from the scene of arrest (ss. 30A to D PACE). This is likely to apply to less serious offences only. In more serious or complex cases the decision to bail usually needs the time to research all the surrounding circumstances of the offence and offender. The withholding of the right to bail can only be justified where certain exceptions apply.

Exceptions to Bail

All accused people (except those mentioned in s. 25 of the 1994 Act) have a right to bail. There will be occasions though, when bail will not be appropriate. In other words, there are exceptions to bail. These exceptions will depend on the individual circumstances. Try the next activity which will lead you through the exceptions.

Albert has been charged with theft and deceptions and asked for bail. The custody officer has decided to keep Albert in custody until the next court sitting. Consider what you know already and the antecedent information that follows. Put yourself in the role of the CPS advocate dealing with Albert's first appearance at court. With regard to the granting of bail by the magistrates, list the key information the advocate will need to bring to the attention of the bench.

Notes:

....

....

....

....

....

....

....

....

....

....

....

....

Westshire Constabulary

Antecedents

List of previous convictions attached.   CRO No ....

1. Full Name ......Albert TURNER...

2. Address ......7 Fyne Road, Porthill, Sandford...

3. Age ..24.....   4. Date of Birth ..23.4......   5. Place of Birth ...Gateshead...........

6. Nationality ...British.................   7. Date of 1st entry into UK.......................

8. Occupation ..Unemployed..

9. Education ....Received full-time education to age 16 yrs attending Eldene.............

Comprehensive School (1 CSE woodwork)...

10. Home Conditions, Domestic Circumstances and Financial Commitments

Separated from wife, having no children, living in furnished bedsit, paying £27.50pw rent.......

Pays nothing towards upkeep of wife..

11. Main Employments during last 5 years (show dates, places and positions held, reasons for leaving and in the case of present employment – wages).

June 19 . . to January 19 . . Apprentice Paint Sprayer, Sandford, dismissed for dishonesty.......

January 19 . . to March 19 . . Shelf Filler, Tesco, Sandford, left of own accord...................

Unemployed to date..

12. Outstanding Matters (e.g. breach of suspended sentence, probation order, conditional discharge, on licence from prison, on bail for other offences, 'totting up, driving disqualification). ....Current offence committed while on bail to Sandford Magistrates' Court for............

offence of theft from off licence...

13. Any other useful antecedent information.

2 burglaries re conviction on 2.8. at Sandford Crown Court were committed while on bail to Sandford.....

Magistrates' Court..

Previous Convictions

Previous Convictions					
Date	Court	Offence(s) with details of any offence taken into consideration	Sentence	Date of release	Show if spent
20.1	Sandford Magistrates' Court	Theft of spray gun (value £40)	Fine £100 Orders to pay £40 compensation		
18.3	Sandford Magistrates' Court	1. Burglary 2. Burglary	1 & 2 Community rehabilitation order for 2 years		
3.3	Sandford Crown Court	1. Burglary 2. Burglary 3. TDA	1, 2 & 3 Borstal Training. Community rehabilitation order put aside	3.9	
6.7	Sandford Magistrates' Court	Making off without payment. (Meal at restaurant)	Fine £50 Ordered to pay £4.50 compensation		
2.8	Sandford Crown Court	1. Burglary 2. Burglary 3. Burglary 4. Failing to appear 10 TICs for burglary	1. 18 months' imprisonment 2. 18 months' imprisonment concurrent. 3. 18 months' imprisonment concurrent. 4. 3 months' imprisonment consecutive. (21 months)	7.9	

Your notes could look something like this:

Notes:

1. He is likely to fail to surrender to custody because:
    (a) of the serious nature of offences (£3,000 loss),
    (b) he is likely to get custodial sentence if found guilty,
    (c) he lives alone in a bedsit (separated from wife).
    (d) he has a long list of previous convictions,
    (e) he has failed to surrender to bail in the past,
    (f) there is a lot of evidence against Albert.

2. He is likely to commit offences while on bail because:
    (a) of the long list of previous convictions,
    (b) he is unemployed,
    (c) he has committed offences while on bail previously.

You may be wondering why we wrote these notes in this way. They actually come straight from the Bail Act 1976, as amended by the Criminal Justice and Public Order Act 1994. There may be others that you thought of also. Let us see what the Act says about exceptions to bail.

The chart below lays out some of the exceptions to bail. There are four important rules, which the court will use regularly. They are in the numbered Exceptions 1, 2, 3 and 4. When deciding if an exception applies to an accused person, the magistrates must take into account the considerations shown in the boxes (A), (B), (C) or (D) in respect of each exception, plus any other consideration, which appears relevant to the court. So these considerations, although for the court, are a clear guide to you when providing information to the CPS. However, it is important to remember that they relate to offences punishable with imprisonment only (sch. 1, part 1, Bail Act 1976).

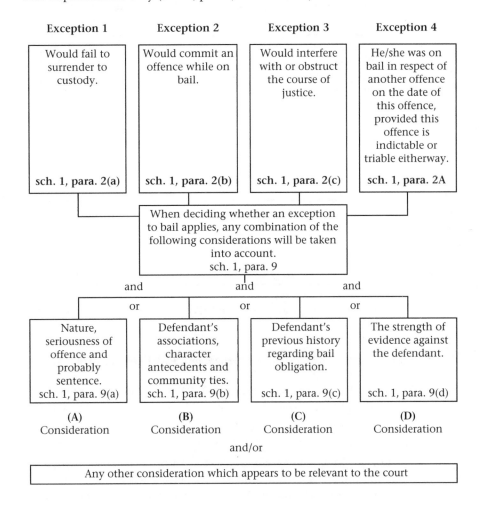

Here are the exceptions as they appear in the Act.

Schedule 1, part 1, Bail Act 1976

2. The defendant need not be granted bail if the court is satisfied that there are substantial grounds for believing that the defendant, if released on bail (whether subject to conditions or not) would—
   (a) fail to surrender to custody, or
   (b) commit an offence while on bail, or
   (c) interfere with witnesses or otherwise obstruct the course of justice, whether in relation to himself or any other person.

2A. The defendant need not be granted bail if—
   (a) the offence charged is an indictable offence or an offence triable either way; and
   (b) it appears to the court that he was on bail in criminal proceedings on the date of the offence.

2B. The defendant need not be granted bail in connection with extradition proceedings if—
   (a) the conduct constituting the offence would, if carried out by the defendant in England and Wales, constitute an indictable offence or an offence triable either way; and
   (b) it appears to the court that the defendant was on bail on the date of the offence.

There are four further exceptions that the court can use, if bail is to be refused. Look at the diagram below which simplifies these exceptions.

Exceptions

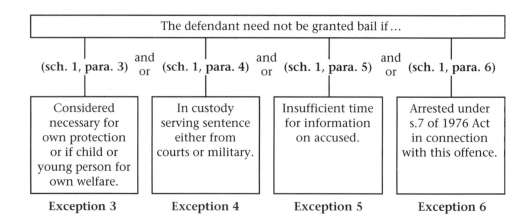

The defendant need not be granted bail if...			
(sch. 1, para. 3)	(sch. 1, para. 4)	(sch. 1, para. 5)	(sch. 1, para. 6)
Considered necessary for own protection or if child or young person for own welfare.	In custody serving sentence either from courts or military.	Insufficient time for information on accused.	Arrested under s.7 of 1976 Act in connection with this offence.
Exception 3	Exception 4	Exception 5	Exception 6

and or between columns.

Do any of these exceptions apply in Albert's case? Tick the answer(s) that apply to him.

(a) Exception 3 ☐   (c) Exception 5 ☐

(b) Exception 4 ☐   (d) Exception 6 ☐

None of the exceptions apply in this particular case. Here they are in full.

Schedule 1, Part 1 Bail Act 1976

3. The defendant need not be granted bail if the court is satisfied that the defendant should be kept in custody for his own protection, or if he is a child or young person, for his own welfare.

4. The defendant need not be granted bail if he is in custody, in pursuance of the sentence of a court or of any authority acting under any of the Services Acts.

5. The defendant need not be granted bail where the court is satisfied that it has not been practicable to obtain sufficient information for the purpose of taking the decisions required by this part of this schedule for want of time since the institution of the proceedings against him.

6. The defendant need not be granted bail if, having been released on bail in, or in connection with, the proceedings for the offence, he has been arrested in pursuance of section 7 of this Act.

It is intended that exceptions to bail be presented to the court by the Crown Prosecution Service. Therefore, it is essential that all the relevant information is supplied in a presentable form to enable the advocate to make proper representation.

These eight exceptions only apply to imprisonable offences, and as far as *Exception 4* is concerned the offence has to be indictable or triable either way as well.

Conditions of Bail

If the court is satisfied that bail can be granted, it may attach conditions to the grant of bail. Under the Bail Act 1976 there are only three conditions that can be attached to bail. All three can be used by the court, and by the police, except for the purpose of medical reports and interviews with advocates. Have a look at the diagram below which sets them out.

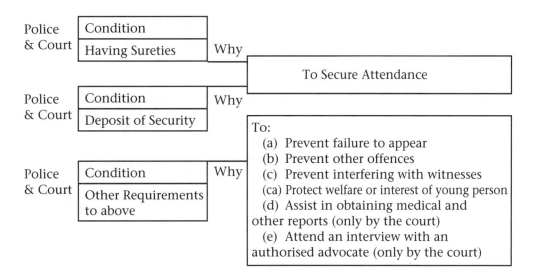

Imagine that in our case, the custody officer had been thinking of giving Albert bail, but is unsure he will turn up at court. Tick from the list those things that the custody officer can do in order to ensure Albert attends court.

(a) Take a deposit of security of £100 cash ☐

(b) Seize his passport ☐

(c) Accept two sureties of £50 each ☐

(d) Insist that he resides with his mother in Wales ☐

(e) Insist that he does not leave his place of residence between 10 pm and 6 am ☐

(f) Insist that he does not come within 1/2 mile of the victim's house ☐

(g) Notify change of address ☐

(h) Report periodically to local police station ☐

(i) Insist that he does not contact victim or witnesses ☐

The custody officer can do all the above to ensure Albert attends court.

The security may be required only from a person who appears unlikely to remain in Great Britain until the time appointed for surrender.

Section 3, Bail Act 1976

The Bail Act 1976 sets out three types of conditions which may be attached to the granting of bail in criminal proceedings at a police station and court.

Court and Police Station

1. Provide surety or sureties.

2. Deposit of security, if it appears that the accused is unlikely to remain in Great Britain.

3. Impose other requirements to be complied with, either before or after release.

The police and the court may apply requirements to the bail of the accused, e.g. seizure of passport, curfew, etc. These requirements can be for one or more of the following reasons only: police s. 3(6)(a), (b), and/or (c); courts s. 3(6)(a), (b), (c) and/or (d), (e).

Section 3(6), Bail Act 1976

He may be required to comply, before release on bail or later, with such requirements as appear to the court [or police, see s. 3A] to be necessary to secure that—

(a) he surrenders to custody,

(b) he does not commit an offence while on bail,

(c) he does not interfere with witnesses or otherwise obstruct the course of justice whether in relation to himself or any other person,

(d) he makes himself available for the purpose of enabling inquiries or a report to be made to assist the court in dealing with him for the offence.

(e) before the time appointed for him to surrender to custody, he attends an interview with an authorised advocate . . .

As investigating officer, it is important that you are aware of these rules and again pass *all* relevant information to the custody officer and the CPS. At any time the magistrates can ask the advocate for information to assist them in making decisions regarding bail.

Under the Police and Criminal Evidence Act 1984, the custody officer can apply conditions to a defendant's bail from the police station. It must be stressed that this is to be applied only to bail to court *not* to return to the police station.

Arrest

There is a power of arrest under s. 7, Bail Act 1976. It enables a police officer to arrest someone who has been released on bail, provided the officer is satisfied about certain points.

Let us assume Albert was granted bail at court and that his wife was accepted as surety for £500. Look at the events that took place on the evening after Albert had been given bail. Which is the earliest time you could lawfully arrest Albert without warrant? Tick the appropriate box.

(a) At 6 pm you receive reliable information from Albert's friend that he is likely to flee the country. ☐

(b) At 7 pm his wife telephones you to say that she believes Albert will abscond. ☐

(c) At 8 pm Albert packs his bags and calls a mini-cab to take him to the airport. ☐

(d) At 9 pm his wife serves written notice on you that she believes Albert will abscond. ☐

The answer is, as soon as you receive reliable information from Albert's friend that he is likely to flee the country, you can arrest Albert. You have reasonable grounds for believing that Albert is not likely to surrender to custody. Look at your powers under s. 7. You will see that s. 7(3)(a) applies in this case.

Section 7(3), Bail Act 1976

A person who has been released on bail in criminal proceedings and is under a duty to surrender into the custody of a court may be arrested without warrant by a constable—

(a) if the constable has reasonable grounds for believing that that person is not likely to surrender to custody;

(b) if the constable has reasonable grounds for believing that that person is likely to break any of the conditions of his bail or has reasonable grounds for suspecting that that person has broken any of those conditions; or

(c) in a case where that person was released on bail with one or more surety or sureties, if a surety notifies a constable in writing that that person is unlikely to surrender to custody and for that reason the surety wishes to be relieved of his obligations as a surety.

Notice that when a surety wants to withdraw, it must be in writing and can only be for the reason that the accused is unlikely to surrender to custody.

Under s. 5B, Bail Act 1976, where bail has been granted by the police or a court, the prosecutor may apply to the appropriate court:

(a) for the conditions to be varied;

(b) to improve conditions where unconditional bail was granted;

(c) to withhold bail.

This only applies to indictable offences or to offences triable either way and will occur only when new information comes to light which was not available at the time the decision was originally made, more often than not prior to the next scheduled appearance in court (e.g. witnesses come forward and complain about intimidation by the accused).

If the person bailed is not present at this hearing, the court may order surrender forthwith into custody. If the bailed person fails to do so, there is a power of arrest under s. 5B(7).

Section 5B(7), Bail Act 1976

A person who has been ordered to surrender to custody under subsection (5) above may be arrested without warrant by a constable if he fails without reasonable cause to surrender to custody in accordance with the order.

The bailed person does not commit an offence but should be brought before the court in the area in which he or she was arrested as soon as practicable, and in any event within 24 hours (s. 5B(8)).

Failure to Answer Bail to Police Station

Although bail to police stations has not been specifically mentioned here, consider the following situation:

What if, instead of being charged with the offence, Albert had been bailed to return to the police station in two weeks time for an ID parade.

Albert does not attend at the given time. What could you now do?

(a) Arrest him for the original offence. ☐

(b) Nothing as no new evidence had come to light. ☐

(c) Write a letter rearranging the time and date for the parade. ☐

(d) Arrest Albert for his failure to attend at the police station. ☐

The answer is (d).

Under s. 46A, Police and Criminal Evidence Act 1984, the police have a power to arrest without warrant someone in Albert's position.

This section does not create a new offence but imposes a specific power of arrest.

Summary

Every accused person (other than those exceptions mentioned) has the right to bail, except where circumstances surrounding the case dictate otherwise. Bail may be refused, granted unconditionally or with conditions attached to the bail.

Self-assessment Test

Having completed this section, test yourself against the objectives outlined at the beginning of the section. You will find the answers below.

1. There is a general right to bail in criminal proceedings (and in particular shall be granted unconditionally or conditionally). With regard to the Bail Act 1976 and s. 25, Criminal Justice and Public Order Act 1994, which of the following is correct?

    (a) George was charged with grievous bodily harm and had a previous conviction for rape. George must not be granted bail in these circumstances.

    (b) Norma was charged with manslaughter and had no previous convictions. Norma has a young child who has special needs. Norma can be granted bail only if this is considered exceptional circumstances.

    (c) Dean was charged with rape and had a previous conviction for attempted rape. Dean could be granted bail only if there were exceptional circumstances.

    (d) Joe was charged with rape and had a previous conviction for manslaughter. Joe had been sentenced to three years' imprisonment for the manslaughter. Joe could not be granted bail even if there were exceptional circumstances.

2. With regard to the Bail Act 1976, which of the following is *not* an exception to bail? If released on bail the defendant would:

    (a) Fail to make himself available for the purpose of enabling inquiries or a report to be made to assist the court in dealing with him for the offence.

    (b) Commit an offence while on bail.

    (c) Interfere with witnesses or otherwise obstruct the course of justice, whether in relation to himself or any other person.

    (d) Fail to surrender to custody.

3. With regard to the three types of condition that can be attached to the granting of bail by a court or custody officer, which of the following is *not* a type of condition?

    (a) Deposit a security of cash.

    (b) Impose other requirements to be complied with, either before or after release.

    (c) Undertaking not to commit further offences while on bail.

    (d) Provide surety or sureties.

Answers to Self-assessment Test

1. Answer (c). This is the only set of circumstances described that meets all the requirements of s. 25.

2. Answer (a). This is one of the conditions that the court can apply when granting bail.

3. Answer (c). This is not one of the three types of condition.

Disclosure

Objectives

With regard to disclosure and the Criminal Procedure and Investigations Act 1996, at the end of this section, from a given set of circumstances, you will be able to:

1. Identify the two tests that the CPS applies when deciding to prosecute.
2. Identify 'relevant' material.
3. Distinguish between unused non-sensitive and sensitive material.
4. Identify where in the case papers primary disclosure is reported to the CPS.
5. Identify the three ways a public interest immunity is decided.
6. Identify when a secondary disclosure is reported to the CPS.

The Role of the CPS in Deciding to Prosecute

Let us look at the role of the Crown Prosecution Service (CPS) in deciding whether to bring a person to court. The CPS are guided by the Code for Crown Prosecutors.

The Code states that the CPS must be satisfied that a case passes two tests before allowing it to proceed:

(a) there is enough evidence;

(b) the public interest requires a prosecution.

These are sometimes referred to as 'the evidential test' and 'the public interest test'.

So there are two stages in the decision to prosecute. The first stage is the evidential test. If the case does not pass the evidential test, it must not go ahead, no matter how important or serious it may be.

If the case does pass the evidential test, a Crown Prosecutor must decide if a prosecution is needed in the public interest. The CPS will start or continue a prosecution only when a case has passed both tests.

We will now consider those tests in more detail.

The Evidential Test

What does sufficient evidence actually mean?

Crown Prosecutors must be satisfied that there is enough evidence which can be used and that the evidence is reliable to show that a criminal offence has been committed by the defendant so as to establish a realistic prospect of conviction.

What constitutes a 'realistic prospect of conviction'? This means that a court is more likely than not to convict the defendant of the charge alleged.

Can the evidence be used? Legal rules may mean that not every piece of evidence can be used in court. You will know, for example, that hearsay evidence is likely to be excluded.

Can you think of any other circumstances in which evidence is likely to be excluded? You will probably have thought about evidence obtained in breach of the Police and Criminal Evidence Act 1984. Details of the defendant's previous convictions are generally inadmissible.

Is the evidence reliable?

Can you think of any factors which would make evidence unreliable?

....

....

....

....

You may have considered some of these:

(a) Is it likely that a confession is unreliable, for example, because of the defendant's age, intelligence, or lack of understanding?

(b) Does the witness have a dubious motive? Is the witness exaggerating? Does the witness have a previous conviction for dishonesty or perjury?

(c) Is the evidence about the defendant's identity strong enough?

(d) Do eyewitnesses contradict each other?

(e) Is the evidence of the eyewitness so similar as to suggest that they have concocted their stories?

So that Crown Prosecutors can make informed decisions about the strength of the evidence, the file should contain all the relevant information.

You are likely to know best, for example, what impression a witness is likely to make. Tell the Crown Prosecutor. Any information which has influenced the decision to charge should be given to the Crown Prosecutor.

The Public Interest Test

Does enough evidence mean that prosecution will be automatic?

Where there is enough evidence, the public interest must be considered. We have already seen that there are two stages in the decision to prosecute—the evidential test and the public interest test.

In 1951, Lord Shawcross, who was Attorney-General, said:

> It has never been the rule in this country—I hope it never will be—that suspected criminal offences must automatically be the subject of prosecution.

This classic statement on public interest has been supported by Attorney-Generals ever since.

In cases of any seriousness, a prosecution will usually take place, unless there are public interest factors against prosecution, which clearly outweigh those tending in favour.

List some public interest factors in favour of prosecution.

....

....

....

....

....

....

....

You probably identified some of these:

(a)   a conviction is likely to result in a significant sentence;

(b)   a weapon was used or violence was threatened during the commission of the offence;

(c)   the offence was committed against a person serving the public (for example, a police officer, a prison officer or a nurse);

(d)   the defendant was in a position of authority or trust;

(e)   the evidence shows that the defendant was a ringleader or an organiser of the offence;

(f)   there is evidence that the offence was premeditated;

(g)   there is evidence that the offence was carried out by a group;

(h)   the victim of the offence was vulnerable, has been put in considerable fear, or suffered personal attack, damage or disturbance;

(i)   the offence was motivated by any form of discrimination against the victim's ethnic or national origin, sex, religious beliefs, political views or sexual preference;

(j)   there is a marked difference between the actual or mental ages of the defendant and the victim, or if there is any element of corruption;

(k)   the defendant's previous convictions or cautions are relevant to the present offence;

(l)   the defendant is alleged to have committed the offence while under an order of the court;

(m)   there are grounds for believing that the offence is likely to be continued or repeated, for example, by a history of recurring conduct; or

(n)   the offence, although not serious in itself, is widespread in the area where it was committed.

Now list some public interest factors against prosecution.

....

....

....

....

....

You are likely to have picked out some of these:

(a) the court is likely to impose a very small or nominal penalty;

(b) the offence was committed as a result of a genuine mistake or misunderstanding (these factors must be balanced against the seriousness of the offence);

(c) the loss or harm can be described as minor and was the result of a single incident, particularly if it was caused by a misjudgment;

(d) there has been a long delay between the offence taking place and the date of the trial, unless:

   (i) the offence is serious,

   (ii) the delay has been caused in part by the defendant,

   (iii) the complexity of the offence has meant that there has been a long investigation;

(e) a prosecution is likely to have a very bad effect on the victim's physical or mental health, always bearing in mind the seriousness of the offence;

(f) the defendant is elderly or is, or was at the time of the offence, suffering from significant mental or physical ill-health, unless the offence is serious or there is a real possibility that it may be repeated;

(g) the defendant has put right the loss or harm that was caused (but defendants must not avoid prosecution simply because they can pay compensation); or

(h) details may be made public that could harm sources of information, international relations or national security.

What about Victims of Crime?

The interests of the victim are an important factor to be carefully taken into account when deciding where the public interest lies. But the CPS acts in the public interest, not just in the interests of any one individual.

What if public interest factors conflict? Crown Prosecutors make an overall assessment of the case, considering the weight to be attached to each public interest factor. Deciding the public interest is not simply a matter of adding up the number of factors on each side.

You may have information which will help a Crown Prosecutor to make the right decision. For example, your case may involve a minor offence which is likely to attract a nominal penalty only. However, it may be a type of offence which is prevalent in a particular area and causes disproportionate annoyance to residents, or involves aggravating factors, such as racial motivation or committing an offence whilst on bail.

Tell the Crown Prosecutor!

The Criminal Procedure and Investigations Act 1996

Disclosures and Unused Material

Sometimes there will he conflict between the public interest in protecting information and in ensuring a fair trial. We will examine this in more detail below.

All material relevant to an investigation should be recorded and retained. Any of this material that does not form part of the prosecution case is regarded as 'unused'.

What does 'relevant' mean? Material will be 'relevant' whether it is beneficial to the prosecution case, weakens the prosecution case or assists the defence case. It is not only material that will become evidence in the case that should be considered, but any information, record or thing, which *may have a bearing* on the case.

In making their decisions, the CPS need to know about unused material and they will rely on you to tell them. You do so by using the MG6C and MG6D schedules. It is the prosecutor's duty to disclose any information to the defence that might undermine the case for the prosecution. This is carried out in two stages—primary and secondary disclosure.

The disclosure officer (who may be you the officer in the case, another investigator or designated support staff) will classify and enter, unused material as either 'non-sensitive' on a MG6C or 'sensitive' on a MG6D.

Non-sensitive Material (MG6C)

Can you think of some items which are likely to be listed on this schedule?

....

....

....

....

You are likely to have thought of these:

(a) interviews (notes and recording) with witnesses, suspects and defendants;

(b) draft statements;

(c) correspondence and notes;

(d) any descriptions of the offender;

(e) notebooks and incident report books.

Sensitive Material (MG6D)

Can you think of some items which are likely to be listed on this schedule?

....

....

....

....

....

....

You probably listed some of these:

(a) matters relating to national security;

(b) information relating to confidential sources;

(c) intelligence;

(d) observation posts and surveillance;

(e) internal police communications;

(f) police/CPS material;

(g) participants in ID parades.

Disclosure Officer's Report (MG6E)

The disclosure officer will then apply the test for 'primary disclosure' to both schedules. This amounts to identifying unused material that appears to undermine the prosecution case. A record is made of this on the disclosure officer's report MG6E. Any items on this report as well as any first descriptions of the suspect, explanations by the accused and any material casting doubt on the reliability of any confession or witness, should be copied if it is practical and sent to the CPS.

Primary Disclosure

The prosecutor will independently consider the MG6 schedules and decide which material meets the test for primary disclosure (which may involve further discussion between the prosecutor and disclosure officer). The CPS then disclose any non-sensitive material meeting the primary test to the defence by forwarding them a copy of the MG6C.

Withholding Information

If any sensitive material appears to meet the test for primary disclosure, an application for public interest immunity (PII) will need to be made.

There are three types of PII application:

(a) open;

(b) *ex parte* with notice;

(c) secret.

An *open* application is one where both parties are present and may make representations to the judge. This would in practice be an unlikely or rare occurrence.

An *ex parte* with notice application is one where the defence is notified of a prosecution representation but would not be present at the hearing or be aware of the nature of the material upon which the application is based. This is likely to include such matters as covert human intelligence sources and surveillance material.

A *secret* application is one where the defence would not be informed of the application or the material. This is likely to include matters of public security.

Even after PII protection has been granted, if the defence are aware of the existence of material and wish access to it, this can be withheld from the defence only if a judge makes an order to this effect.

Defence Statements

After primary disclosure has taken place, the defence *must* in the case of Crown Court proceedings and *may* in the case of magistrates' court proceedings serve a 'defence statement' on the prosecution. This should include:

(a) the accused's defence;

(b) matters on which there is disagreement with the prosecution;

(c) the reasons for the disagreement.

Secondary Disclosure

The defence statement is passed from the CPS to the disclosure officer who will re-examine the unused material for any items which might assist the defence case. This is secondary disclosure. It is forwarded to CPS on another MG6E. If the material is sensitive, this may trigger another PII hearing.

What if the Judge Orders Disclosure?

If there is a conflict between ensuring a fair trial or withholding material, the judge will always order disclosure. The prosecution will be faced with giving the information to the defence or abandoning the case.

Summary

Because the CPS and police work closely together, it is vital that we know how decisions to prosecute are made. The Code for Crown Prosecutors is a public statement of principles and is available at any police station. The Manual of Guidance for the preparation, processing and submission of files assists in the preparation of files.

Self-assessment Test

Having completed this section, test yourself against the objectives outlined at the beginning of the section. You will find the answers below.

1. With regard to the Crown Prosecution Service, which of the following is correct?
   The Code for Crown Prosecutors states that Crown Prosecutors:
   (a) Must be satisfied that there is enough evidence, which can be used and is reliable, to show that a criminal offence has been committed by the defendant so as to establish a realistic prospect of conviction.
   (b) Can decide to proceed with a prosecution when they are satisfied that there is some evidence and it is reliable to show that a criminal offence has been committed by the defendant so as to establish a probable prospect of conviction.

(c) Must apply the public interest test before they take the evidential test into account when deciding to proceed with a prosecution.

(d) Must apply two evidential tests before they can decide whether to proceed with a prosecution.

2. With regard to the Criminal Procedure and Investigations Act 1996, which of the following correctly completes the definition of 'relevant' material?

Material will be 'relevant' whether it is beneficial to the prosecution case if it,

(a) weakens the prosecution case or assists the defence case;

(b) weakens the prosecution case. It is not only material that will become evidence in the case that should be considered, but any information, record or thing, which *must have a bearing* on the case;

(c) assists the defence case. It is only material that will become evidence in the case that should be considered and any information, record or thing, which will *have a bearing* on the case;

(d) weakens the prosecution case or assists the defence case. It is not only material that will become evidence in the case that should be considered, but any information, record or thing, which *may have a bearing* on the case.

3. With regard to primary disclosure, which of the following would be classified as sensitive material?

(a) Notebooks and incident report books.

(b) Participants in ID parades.

(c) Interview notes from a witness statement.

(d) Any descriptions of the offender.

4. With regard to an application for public interest immunity (PII), which of the following is *not* a type of application?

(a) Open application.

(b) Open application without notice application.

(c) *Ex parte* with notice application.

(d) Secret application.

Answers to Self-assessment Test

1. Answer (a). This is the evidential test to be applied by the CPS.

2. Answer (d). This is the complete definition of 'relevant' material.

3. Answer (b). All the others are non-sensitive material.

4. Answer (b). An open application must by definition be with notice.

Accessories and incomplete offences

Objectives

With regard to accessories and incomplete offences, at the end of this section, from a given set of circumstances, you will be able to:

1. Identify when a person aiding, abetting, counselling or procuring another to commit an offence would be treated as a principal.
2. Identify the points to prove for attempting to commit a criminal offence (s. 1, Criminal Attempts Act 1981).

Introduction

The law accepts that there are people who, though they do not themselves physically commit a crime, will nevertheless be considered as principals relating to the criminal activities of others. The following legislation deals with such cases.

Section 8, Accessories and Abettors Act 1861

Whosoever shall aid, abet, counsel or procure the commission of any indictable offence, whether the same be an offence at common law or by virtue of any Act passed or to be passed, shall be liable to be tried, indicted, and punished as a principal offender.

Section 44(1), Magistrates' Court Act 1980

A person who aids, abets, counsels or procures the commission by another person of a summary offence shall be guilty of the like offence and may be tried (whether or not he is charged as a principal) . . .

Aiding and Abetting

Cassady v Reg Morris (Transport) Ltd (1975)

To aid and abet, a person must, with knowledge, give assistance or encouragement of some kind to the actual perpetrator.

Mere presence will rarely constitute aiding and abetting. It must be proved that the person or persons were giving some assistance or encouragement and that they were acting in concert.

A person acting as a lookout for the person committing the crime is aiding and abetting, if he has knowledge of the offence and is acting in concert, and is near enough to render assistance should the need arise.

Procure

To procure means to produce by endeavour. A person procures an offence by setting out to see that it happens and taking the appropriate steps to produce that happening.

This amounts to persuading or organising others to commit an offence in which the organiser is taking no active part, i.e. the gang boss who sets up a crime, but is not involved in its actual commission. This offence can be charged only when the offence has been committed.

This offence differs from conspiracy in that there is not an agreement of two or more to commit an offence; it is an order or arrangement by one for another to commit the offence.

The procurement of an offence must be continuous. If the procurer changes his or her mind and withdraws the orders, he or she cannot be guilty of this offence.

If a person procures someone to commit an offence, e.g. theft, and that person commits a different offence, e.g. grievous bodily harm, the original person cannot be guilty of procuring grievous bodily harm.

Counselling

This involves giving advice prior to an offence. A person who counsels the commission of an offence is guilty if he or she intends that act of counselling, whether or not he or she desires the offence actually to be committed.

R v Calhaem [1985] QB 808

It is not necessary for the counselling to be a substantial cause of the commission of the offence.

Attempts to Commit Crime

Section 1, Criminal Attempts Act 1981

(1) If, with intent to commit an offence to which this section applies, a person does an act which is more than merely preparatory to the commission of the offence, he is guilty of attempting to commit the offence.

(2) A person may be guilty of attempting to commit an offence to which this section applies even though the facts are such that the commission of the offence is impossible.

(3) In any case where—
    (a) apart from this subsection a person's intention would not be regarded as having amounted to an intent to commit an offence; but
    (b) if the facts of the case had been as he believed them to be, his intention would be so regarded, then, for the purposes of subsection (1) above, he shall be regarded as having had an intent to commit that offence.

(4) This section applies to any offence which, if it were completed, would be triable in England and Wales as an indictable offence, other than—
    (a) conspiracy (at common law or under section 1 of the Criminal Law Act 1977 or any other enactment);
    (b) aiding, abetting, counselling, procuring or suborning the commission of an offence;

(c) offences under section 4(1) (assisting offenders) or 5(1) (accepting or agreeing to accept consideration for not disclosing information about an arrestable offence) of the Criminal Law Act 1967.

This means a person may be guilty of attempting to commit an offence, even though the facts are such that commission of the offence is impossible (e.g. where someone tries to steal from an empty pocket).

More than Merely Preparatory

An intention to commit an offence is not sufficient because criminal liability does not commence until the offender takes some steps towards the commission of the crime.

Preparation for an intended crime will not amount to an attempt, e.g. buying a box of matches will not be attempted arson.

The intent must be shown by some overt or open act connected with the commission of the intended crime.

Whether the defendant has gone beyond the point of 'merely preparatory' will be a question for the jury or magistrate. There is no requirement for the defendant to have passed a point of no return, but he or she need to have 'embarked on the crime proper'.

R v *Tosti* (1997) Crim LR 746

In this case the defendants took oxyacetylene equipment to the scene of a planned burglary and concealed the equipment in a hedge. They approached the door of a barn and examined the padlock on it. At that point they realised they were being watched and ran off. They appealed against their convictions for attempted burglary. It was held, dismissing their appeal, that there was evidence that they had done an act which showed that they had actually attempted to commit the offence. The judge had been correct to leave the issue to the jury.

A person indicted for an offence may be convicted of the attempt to commit that crime, if upon the trial it is shown that he or she attempted, but did not complete it.

Mens Rea in Attempts

It is not generally true that the *mens rea* of an attempt is essentially the same as the *mens rea* of the complete crime. For instance, the offence of assault can be committed intentionally or recklessly. An attempted assault, however, can be committed intentionally only.

Also the *mens rea* for attempted murder is narrower than the *mens rea* for murder. Murder can be committed when the defendant intends either to kill or to commit grievous bodily harm. Attempted murder can be committed when the defendant intends to kill only.

The full rule is that in an attempt the defendant must intend the prohibited result, but need only be reckless as to any relevant circumstance (see *R* v *Millard* [1987] Crim LR 393).

Summary

In this section you have covered offences that are preparatory acts to the crime itself. In the following section you will be looking at the offence of conspiracy.

Self-assessment Test

Having completed this section, test yourself against the objectives outlined at the beginning of the section. You will find the answers below.

1. With regard to the Accessories and Abettors Act 1861, which of the following commits an offence?
   (a) Liam knew that Habib had planned a future robbery, but denied all knowledge of this when questioned by the police.
   (b) Owen supplied a can of petrol to Heather so that she could commit arson, but was not present when she committed the arson.
   (c) Ian walked past a building society and saw a robbery in progress, but did nothing to either prevent it, or to alert the police.
   (d) Kylie knew her boyfriend did not hold a driving licence and allowed him to drive her car on the road.

2. With regard to s. 1, Criminal Attempts Act 1981, at what stage does Barry commit the offence of attempted burglary?
   (a) Barry intended to commit a burglary some time in the future. He drove out to a large country house and looked for any movements of the residents and potential routes in and out of the estate. He returned home to prepare for the burglary.
   (b) Barry then went to a builder's merchant and bought a pair of bolt cutters and industrial gloves. He drove out to the large country house and hid the bolt cutters and industrial gloves in a wood adjacent to the house. He intended to return that night to break into the house.
   (c) Barry drove to the wood at 9 pm and picked up the bolt cutters and industrial gloves. He left his car with the equipment and went to a garage attached to the house where he intended to burgle.
   (d) Barry cut the lock on the garage door. He opened the door and went inside looking for anything of value that was worth stealing. He heard noises coming from the house and ran off without stealing anything.

Answers to Self-assessment Test

1. Answer (b). Owen is aiding Heather in committing the offence of arson.
2. Answer (c). Barry, when going to the garage door has done an act, which is more than merely preparatory.

Conspiracy

Objectives

With regard to conspiracy, at the end of this section, from a given set of circumstances, you will be able to:

1. Identify the points to prove for conspiracy (s. 1, Criminal Law Act 1977 as amended by s. 5(1), Criminal Attempts Act 1981).

2. Identify that where an offence is impossible to commit, a person can be guilty of conspiring to commit that offence.

3. Identify the types of conspiracy that can be committed other than under the Criminal Law Act 1977.

Introduction

What is conspiracy?

You no doubt have your own idea of its meaning. The dictionary tells us that to conspire means to 'combine for an evil purpose', to 'plot', to 'devise' etc.

That is an excellent base from which to start when considering the offence of conspiracy in the context of the criminal law.

To help you build the definition as laid down in law, let us go back to Case Study 3 involving our burglars: Piper, Collier and Pickles.

In the first instance, Piper was made aware that the chief executive of a nationwide chain of grocery stores would be leaving his luxurious house unoccupied during the Bank Holiday period. Piper was determined to burgle the premises, alone, if necessary, but invited Collier and Pickles, to take part.

Was this a conspiracy?

Let us look at the definition.

The Definition of Conspiracy

Section 1(1), Criminal Law Act 1977

... if a person agrees with any other person or persons that a course of conduct shall be pursued which, if the agreement is carried out in accordance with their intentions, either—

(a) will necessarily amount to or involve the commission of any offence or offences by one or more of the parties to the agreement, or

(b) would do so, but for the existence of facts which render the commission of the offence, or any of the offences, impossible,

he is guilty of conspiracy to commit the offence or offences in question.

The most important word in the definition is 'agreement'.

So even at this early stage, we can see that our burglars have not conspired. Piper merely formed an intention and, although he invited the others to take part, there was no acceptance and, therefore, no agreement.

An agreement then, must always exist. By reference to the definition, spend a minute or two listing below what you think are the other points necessary to prove the offence of conspiracy:

....

....

....

....

....

....

Having considered the vital ingredients of the offence, try the following activity:

Indicate after each of the following sets of circumstances whether those mentioned are guilty or not guilty of conspiracy.

1. Piper discusses the proposed burglary with Collier and Pickles. After due consideration, Collier and Pickles decline to take part:

	Guilty	Not guilty
(a) Piper	☐	☐
(b) Collier	☐	☐
(c) Pickles	☐	☐

2. Piper discusses the proposed burglary with Pickles and Collier, who both tell him they need time to make up their minds:

	Guilty	Not guilty
(a) Piper	☐	☐
(b) Collier	☐	☐
(c) Pickles	☐	☐

3. Piper discusses the proposed burglary with Collier and Pickles. Collier accepts his invitation to take part but Pickles declines to do so:

	Guilty	Not guilty
(a) Piper	☐	☐
(b) Collier	☐	☐
(c) Pickles	☐	☐

4. Piper discusses the proposed burglary with Collier and Pickles and all decide to take part:

	Guilty	Not guilty
(a) Piper	☐	☐
(b) Collier	☐	☐
(c) Pickles	☐	☐

In the first activity your answer should have been similar to this:

(a) an agreement;

(b) with at least one other person;

(c) to pursue a course of conduct;

(d) which would lead to the commission of an offence by one or more of the parties to the agreement.

So, in order to prove an offence of conspiracy, there must be evidence of an agreement with at least one other person *and* the agreement must be to pursue a course of conduct, i.e. they must agree to carry out a plan, *and* if carried out, that plan would necessarily amount to or involve the commission of an offence, by one or more of those concerned in the agreement.

Once those points are recognised the answers to the second activity are clear.

Looking at the circumstances outlined in questions 1 and 2, it can be seen that no agreement has been reached. That being so, we need look no further. We know there can be no evidence of conspiracy.

In question 3, Collier has accepted Piper's invitation to take part in the burglary. This can properly be construed as an agreement to pursue a course of conduct, which, if carried out in accordance with their intentions, would necessarily amount to the commission of an offence. Piper and Collier are, therefore, guilty of conspiracy. Pickles, of course, in declining the invitation, has not entered into an agreement and remains not guilty.

In question 4, all three have agreed to commit the offence and are, therefore, guilty of conspiracy.

But what if, having entered into the conspiracy, unbeknown to them the chief executive's house has burned down? Now it would be impossible to carry out their plan. Would they still be guilty of conspiracy?

Section 1(1)(b) of the definition gives us the answer. So as to remain consistent with the Criminal Attempts Act 1981, it is an offence, contrary to s. 1(1), Criminal Law Act 1977, to conspire to commit an offence, which in reality, would be impossible to commit.

A person can be convicted of conspiracy even if the other conspirators are unknown. However, a defendant cannot be convicted of a statutory conspiracy if the *only* other party to the agreement is:

(a) his or her spouse;

(b) a child/children under 10 years of age;

(c) the intended victim.

A husband and wife can both be convicted of a statutory conspiracy if they conspire with a third person.

Common Law and other Statutory Conspiracies

As a general rule, conspiracies to commit all types of offences are contrary to s. 1, Criminal Law Act 1977 (as amended by s. 5(1), Criminal Attempts Act 1981). But, as is so often the case in criminal law, there are exceptions to this rule. Those exceptions are:

(a) Conspiracy to engage in conduct which tends to corrupt public morals *or* outrage public decency contrary to Common Law.

(b) Conspiracy to defraud contrary to s. 12, Criminal Justice Act 1987.

It is not intended to go into detail on these exceptions, but you now know that they exist.

Self-assessment Test

Having completed this section, test yourself against the objectives outlined at the beginning of the section. You will find the answers below.

1. With regard to conspiracy (s. 1, Criminal Law Act 1977), which of the following does *not* disclose an offence of statutory conspiracy?
   (a) Kate agreed with her husband David to go to Marks and Spencer, steal clothes from the displays and take them to the returns counter to get the cash in exchange. On the way to the store, Kate lost her nerve and refused to carry out the plan. David went on to commit the offences of theft and deception.
   (b) Marion and her friend Robbie agreed to extract cocaine from a substance they had bought from a contact in the chemical industry. Unknown to them the substance contained no cocaine and they could not have committed the offence of producing a controlled drug.
   (c) Hugh approached Gavin and asked him whether he was interested in robbing a post office. Gavin was not sure and asked for time to think about it. In the mean time, Hugh asked Chris whether he would drive the getaway car in the proposed robbery. Chris refused but Gavin phoned Hugh to say he was prepared to do the robbery.
   (d) Linda and her husband Roger agreed to blackmail their business partner over some unethical business deals the partner had undertaken. They approached James, who had a grudge against the business partner, and asked him to come in with them in the blackmail. James agreed provided he got half of the proceeds.

Answer to Self-assessment Test

1. Answer (a). A husband and wife can only commit statutory conspiracy when a third party is involved in the agreement.

PART FIVE

The Police and Criminal Evidence Act 1984—Codes of Practice Case Study

Introduction

This section is based on an investigation. You are to be the investigating officer and you will be required to work through the process.

How to Use this Section

You will be asked to make decisions during the investigation, and, as a result of your decisions, the investigation will develop. Your route through the section will be dictated by your choices and you will not be required to read all of the paragraphs.

Each paragraph in this section is numbered. At the end of each piece of work or text, there will be instructions to '**Go to**' the next paragraph of the investigation, indicated by **an arrow**. For ease of reference, alongside each paragraph number, the number of the previous paragraph is included in smaller type.

Some questions will require you to write your answers in the text. Space is provided for this and is indicated by the ✎ symbol.

To assist you in your work, you will be asked to write down the paragraphs that you refer to on the grid sheet, overleaf.

A map of the area is included. Feel free to highlight or mark places or streets that are mentioned in the text.

There is no need to complete this section in one sitting. If you wish to take a break, make a note of the paragraph number and resume when you are ready.

You will find the PACE Codes of Practice, Codes A–F set out in full at the end of the **Investigator's Manual**.

As you work through this section, please write down each paragraph number as you go. This will help you as you progress and also, on completion, enable you to check the route that you have taken.

Start here:

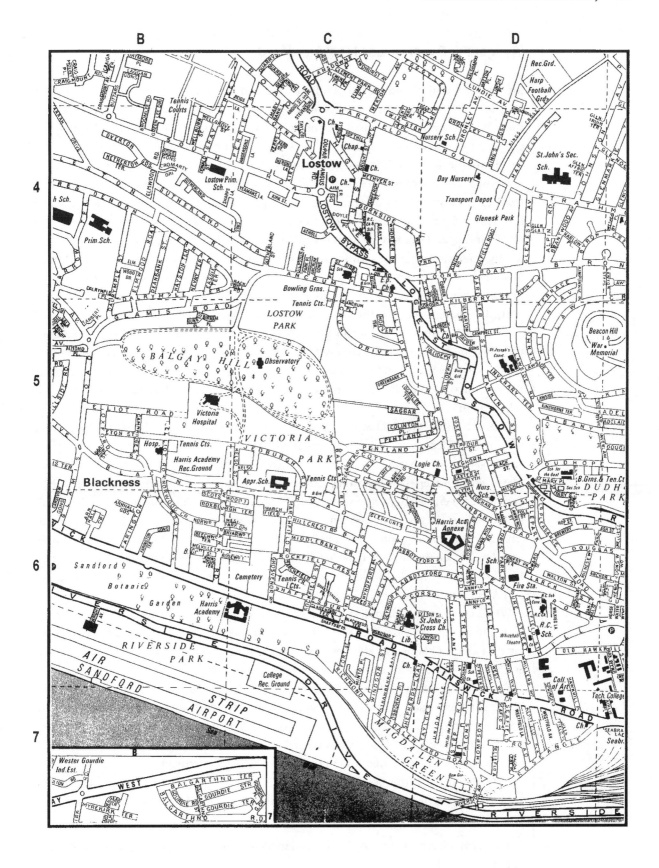

1 | 'Aysgarth', Overton Gardens, Sandford   1

Mr Martin Saunders and his wife Philippa returned to their family home at 3.45 pm this afternoon. Mr Saunders parked the car on the gravel drive and went to the front door. He put the key in the lock and tried to open the door. The key wouldn't work. The lock was apparently set from the inside. Mrs Saunders waited at the front door while her husband went to the side of the house to check the gate to the back garden. Unusually, the gate was unbolted, it was wide open. Mr Saunders ran around the back and saw that the small kitchen windows had been broken and the larger one was wide open. He went to the back door and tried his key. This lock had also been set from the inside. He climbed into the kitchen through the open window and went to the front door. The catch had been set. He unlocked the door and let his wife in. They both went into the sitting room and saw that the room was a mess. Drawers had been emptied and the contents of their display cabinets strewn over the floor. Mrs Saunders started crying. Mr Saunders became angry. Together they went to every room of the house. The story was the same. Mr Saunders called the police.

PC Janet Hopkins was on patrol and was requested to attend. On driving into Overton Gardens, she noticed that the large detached houses were set well back from the road.

After several minutes of looking at the house names, Janet located 'Aysgarth' and parked the car. She walked up the drive and rang the door bell. Mr Saunders opened the door and asked her in.

After the initial introductions, Mr Saunders took Janet on a tour of the house. He pointed out the upturned drawers, emptied cupboards and the general results of an untidy search and began to list what he could see as immediately missing. The mini hi-fi compact disc player and speakers had been taken from the sitting room. All Mrs Saunders' jewellery (a white gold engagement ring set with eight sapphires and eight diamonds, an 18 ct gold chain with an 18 ct gold fob pendant, a 9 ct gold five bar gate bracelet, an 18 ct gold ring set with a single diamond and an 18 ct gold lady's Rolex wrist watch), had been taken from the jewellery box on the dressing table in the main bedroom. The total value was approximately £5,400. Also taken from the bureau in the hallway were the family cheque books and Visa and Access credit cards.

While Janet and Mr Saunders were looking around the house, Mrs Saunders remained seated in the sitting room. She was still very distressed and unable to hold back her tears. Janet Hopkins obtained all the necessary details for the crime report. She requested that SOCO and CID be informed and attend as soon as possible.

The means of entry was through the ground floor kitchen windows at the back of the house. The small double glazed, top window had been smashed with a sharp pointed object in the bottom left-hand corner. The secondary window had then been broken and the larger lower window had been opened. The windows had locks on them but the family never used them.

Janet and Mr Saunders were still in the kitchen when you arrived.

What do you do at this point?

Get the details from the PC and take over completely allowing the PC to go.

►36

Assess what has been done by the PC and work together to complete the necessary scenes of crime enquiries.

►50

With the PC with you, confirm the details of what has happened with the victim and then send the PC away.

►60

99►2 You relate the evidence to the Custody Officer and he is prepared to accept a charge of burglary.   2

CUSTODY RECORD	
Time	**Full details of any action/occurrence involving detained person (Include full particulars of all visitors/officers)** **Individual entries need not be restricted to one line** **All entries to be signed by the writer (include rank and number)**
20.30	Subject charged, property returned and released into his mother's custody. A West Ps

After completing the necessary paperwork in the office, you and a colleague go to the location of the outstanding property and find that the lock-up garage is wide open and there is nothing in it. You check Peter Grey's home address and are informed by a neighbour that he, and a boy called Wayne, recently left in his car.

In hindsight, was charging immediately the best option?

►99
make another choice

51►3 Code C, para. 11.13 and note 11E outlines what you should do. By complying with the   3
instructions, you give the suspect the opportunity to comment on your notes which can be a useful way of introducing the previous conversation into the interview, as described by Code C, para. 11.24.

When you arrive at the Police Station you present your prisoner to Sgt Allan West, the Custody Officer.

What do you tell the Custody Officer when you present the suspect to him?

Does the prisoner have to be present when you talk to the Custody Officer?

....

....

....

....

....

►30

WESTSHIRE CONSTABULARY

4

WITNESS STATEMENT
(C.J. Act. 1967, S.9; M.C. Act 1980,
S.102; M.C. Rules. 1980. r.70)

STATEMENT OF ...... Martin Saunders ....

AGE ...... Over 18 years ....

This statement (consisting of . . . 1 . . . page, each signed by me) is true to the best of my knowledge and belief and I make it knowing that if it is tendered in evidence, I shall be liable to prosecution if I have stated in it anything which I know to be false or do not believe to be true.

DATED ...... 19th day of February ....

SIGNED ...... M Saunders ....

Further to my earlier statement, at 10 pm today, I attended Sandford Police Station where I again spoke to Detective Constable Lee. She showed me a Hitachi mini hi-fi compact disc player with speakers (serial number H/S 129/8793/3645), Visa and Access credit cards in the name M SAUNDERS; Nat West cheque book, name of M & P Saunders, a white gold engagement ring set with eight sapphires and eight diamonds, an 18ct gold chain with an 18ct gold fob pendant (value £500.00); a 9ct gold five bar gate bracelet and an 18ct gold lady's Rolex wrist watch, which I can positively identify as the property stolen from my house earlier today.

SIGNED ...... M Saunders ....

(*continued*) 4 | # WESTSHIRE CONSTABULARY | 4

WITNESS STATEMENT
(C.J. Act. 1967, S.9; M.C. Act 1980,
S.102; M.C. Rules, 1980. r.70)

STATEMENT OF ...... David Attwood ....

AGE ...... Over 18 years ....

This statement (consisting of . . . 2 . . . pages, each signed by me) is true to the best of my knowledge and belief and I make it knowing that if it is tendered in evidence, I shall be liable to prosecution if I have stated in it anything which I know to be false or do not believe to be true.

DATED ...... 19th day of February ....

SIGNED ...... D Attwood ....

I live at the address overleaf. The premises are detached and set in their own grounds.

At 1 pm on Thursday the 18th of February, I locked up the house and went to visit relatives. The house was left unoccupied.

At 8 pm the same day, I returned to my house. I went to the front door and tried to unlock the door with my key. The key would not unlock the door. The catch appeared to have been locked from the inside. I went around the back and noticed that the large dining room window was wide open. I also noticed that the small dining room window adjacent to the window had been broken.

I went to the back door and tried to open it. It also seemed to be locked from the inside. I got a neighbour to climb in through the large window and let me in.

I went into the sitting room and saw that the place had been ransacked.

I went into all the rooms in the house and found that each had been similarly treated. I noticed that the following property had been taken: my Access credit card, number 2009 4773 2235 and my Nat West cheque card, number 3476436X9 had been taken from my jacket which was hanging in the hallway, a yellow gold wedding band engraved with DA 14/11/52 (value £250.00), and some assorted costume jewellery, (value £50.00) had been taken from my bedside cabinet.

I called the police and reported what had happened.

Nobody had any right or authority to enter my house without my permission or to take property belonging to me.

At 11 pm the same day, I attended Sandford Police Station where I was introduced to Detective Constable Lee. She showed me an Access credit card (AA/2), a yellow gold wedding band (AA/3), and various items of costume jewellery (AA/4) that I positively identified as that taken from my house.

I am willing to attend court if required.

SIGNED ...... D Attwood ....

▶91

$^{19}_{63}$▶5   During the afternoon of Sunday the 21st of February, you have been liaising with the Identification Officer, Inspector Atkinson, in order to arrange the parade for 6.30 pm tonight.

You and the Identification Officer had agreed that an identification parade was the most suitable procedure. What other options were available to you?

....

....

....

....

▶18

92▶6   There is a review due before the morning and Inspector Atkinson, the Duty Inspector, has indicated that she is not happy with Peter being kept in custody when the identification procedure could be arranged later, as long as it was as soon as practicable. She tells you that she would not authorise further detention.

This leaves you with no alternative but to bail Peter to come back to the police station at a time when you have arranged the identification procedure.

▶19

74▶7   Code C, para. 3.1, outlines the rights that every suspect has when they come into police detention.

Detained person, Wayne Cooper, is given his rights by the Custody Officer, verbally and written.

As he is a juvenile, what do the Codes of Practice say the Custody Officer must do?

....

....

....

....

▶102

26▶8 There are no right or wrong answers as to what you should tell Mr Edwards, as long as you do not give deliberately misleading information. Remember, you are not obliged to disclose any of the evidence to the solicitor or prisoner at this stage.

You should think carefully about what you do disclose and plan your interview accordingly. You may decide several interviews are needed and plan to disclose new evidence at each stage before each interview.

CUSTODY RECORD	
Time	Full details of any action/occurrence involving detained person (Include full particulars of all visitors/officers) Individual entries need not be restricted to one line All entries to be signed by the writer (include rank and number)
19.15	Custody to D.C......... for the purpose of interview in the presence of his mother and solicitor. a. West. Ps.
19.40	Interview concluded, P.A.C.E and the Codes of Practice complied with. Subject returned to Detention Room.

It is essential that you inform the Custody Officer that you have completed the interview, that PACE and the Codes of Practice have been complied with and the custody record is endorsed to that effect.

The main points to come out from the interview are:

- A full admission of Mr Saunders' burglary, corroborated by knowledge of premises, method of entry, property stolen.

- Name of accomplice—Peter Grey, of Patons Lane, (Map Ref D7).

- Location of the outstanding property in Peter Grey's lock-up garage in Thompson Street (Map Ref D7).

- He is not in possession of any of the stolen property.

- No other admissions.

Having obtained an admission during the interview, what does PACE say about confessions?

....

....

....

....

....

121►9

There is a review due before the morning and Inspector Atkinson, the Duty Inspector has indicated that she is not happy with Peter being kept in custody when the identification procedure could be arranged later, as long as it was as soon as practicable. She tells you that she would not authorise further detention.

This leaves you with no alternative but to bail Peter to come back to the Police Station at a time when you have arranged the identification procedure.

►57

9

83►10

Section 17(2)(a) PACE explains that these powers are only exercisable if you have reasonable grounds for believing that the person who you are seeking is on the premises. In these circumstances, you have seen the youth run into the house and therefore, satisfy the conditions.

If the door had slammed closed behind the youth, what power would you have to enter by force?

....

....

►108

10

73►11

Code A, paras 3.8 to 3.11 outline what you should say.

►112

11

98►12

Code C, Annex B outline the circumstances, however, there are some other conditions that need to apply.

What is a 'serious arrestable offence' and how do the circumstances of this case fit the requirements of Annex B?

....

....

....

....

....

►90

12

106►13

What does PACE say about what you should do when you have evidence of further offences?

....

....

....

►81

13

100► 14

Code D, para. 3.2 outlines the rules that must be adhered to. Note that this paragraph only applies when the suspect is not known.

A possible suspect has been found by a uniformed foot patrol in the Town Centre. Mrs Woodhams is taken to the location and identifies the suspect as the youth who is being spoken to by the uniformed officer.

14

How does this affect the identification?

....

....

....

....

....

....

►45

61► 15

15

Previous Convictions

Convictions recorded against: .... Wayne Cooper .... CRO No ....

Charged in name of .... Wayne Cooper .... Age ...16 yrs....

Date	Court	Offence(s) (With details of any offence taken into consideration)	Sentence	Date of release
Last Year	Sandford Juv	Burglary	Con discharge, 12 months Comp £20.00	
Last Year	Sandford Juv	Burglary Theft of Cycle	Sup Order 2 years Sup Order 2 years Comp £70.00	

►34

111► **16**   Section 19 PACE and Code B, para. 7.1 outline your powers to seize property found during a search.

16

Throughout your search of the house, the suspect, Peter Grey, is with you. What questions can you ask?

....

....

....

....

►62

131► **17**   The solicitor, Mr Edwards (the Duty Solicitor), arrives at the station and is informed by the Custody Officer that he has already advised the co-defendant in this case.

17

What do the Codes of Practice say about solicitors advising co-defendants?

....

....

....

....

....

►110

$^5_{47}$► **18**   The Codes of Practice, Code D, paras 3.4 to 3.10, outline the alternatives of video, identification parade and group identifications. It is appreciated that the person responsible for running these will be the Identification Officer, but as you are the officer in the case it is important to liaise closely with her. However, you are not to take any active part in the parade itself.

18

The parade is held soon after 6.30 pm and Mrs Woodhams positively identifies Peter Grey as the person seen in Overton Gardens two days earlier.

What do you do next?

Charge him with the two burglaries   ►42
Re-interview, re: both offences   ►68

$^6_{92}$► **19**   Peter is bailed to return in two days' time at 6 pm. You are on 2 pm to 10 pm which allows you time to do the final arrangements before he arrives.

19

►5

64► **20**

20

Under the circumstances that you have here, you have no immediate power to enter and search. You would need to obtain a search warrant. Under what circumstances would you not need a search warrant?

....

....

....

....

►77

65► **21** Section 1(6) PACE gives you the power to seize and retain any property found.

21

Having completed the search what must you do next?

....

....

....

....

....

►118

WESTSHIRE CONSTABULARY

WITNESS STATEMENT

(C.J. Act, 1967, S.9; M.C. Act 1980,
S.102; M.C. Rules, 1980, r.70)

STATEMENT OF ....Martin Saunders...

AGE ....Over 18 years...

This statement (consisting of . . . 2 . . . page, each signed by me) is true to the best of my knowledge and belief and I make it knowing that if it is tendered in evidence, I shall be liable to prosecution if I have stated in it anything which I know to be false or do not believe to be true.

DATED ....19th day of February...

SIGNED ....Martin Saunders..

I live at the address overleaf. The premises are detached and set in their own grounds.

At 3.15 pm on Friday the 19th February, I locked up the house and together with my wife, went into town. The house was left unoccupied.

At 3.45 pm the same afternoon. I returned to my house. I went to the front door and tried to unlock the door with my key. The key would not unlock the door. It appeared that the catch had been locked from the inside. I told my wife to stay in the car on the drive and I went to the side gate. This gate is normally kept bolted hut it was wide open. I went around the back and noticed that the large kitchen window was wide open. I also noticed that the small kitchen window had been broken.

I went to the back door and tried to open it. It also seemed to be locked from the inside. I had to climb in through the kitchen window.

I went to the front door and let my wife in. We went into the sitting room and saw that the place had been turned upside down. The drawers had been emptied and the contents of the display cabinets were strewn all over the floor.

We went into all the rooms in the house and found that each had been similarly treated. I noticed that the following property had been taken: a Hitachi mini hi-fi compact disc player with speakers (serial number H/S129/08793/3645) value £400.00, taken from the downstairs sitting room; Visa and Access credit cards in the name M SAUNDERS; Nat West cheque book, name of M & P Saunders, from the bureau in the downstairs hallway; a white gold engagement ring set with eight sapphires and eight diamonds (value £1,500.00); an 18 ct gold chain with an 18 ct gold fob pendant (value £500.00); a 9ct gold five bar gate bracelet (value £250.00); an 18ct gold ring set with a single diamond (value £750.00); an 18ct gold lady's Rolex wrist watch (value £2,000.00), taken from the jewellery box on the dressing table in the main bedroom.

I called the police and reported what had happened.

Nobody had any right or authority to enter my house without my permission or to take property belonging to me.

At 6 pm the same day, I attended Sandford Police Station where I was introduced to Detective Constable Lee. She showed me an 18 ct gold ring set with a single diamond (exhibit AA/1). I can positively identify this ring as one of those taken from my house earlier today.

I am willing to attend court if required.

SIGNED ....Martin Saunders..

▶29

57►**23** You only have circumstantial evidence to suggest that Wayne may have been involved in the other burglary.

    You interview him in the presence of his mother and he admits nothing. You return him to the detention room and consider searching his home in relation to outstanding property from Sutherland Crescent.

►41

122►**24**

What do the Codes of Practice say about conducting a 'voluntary search'?

....

....

....

►120

102►**25** Code C, para. 1.7 outlines what is meant by an appropriate adult and Notes 1B, 1C, 1E, and 1F explain certain extra considerations.

    After being given his rights, Wayne asks for the Duty Solicitor and for his friend Graham Jones to be informed of his detention. These requests are noted in the relevant section of the custody record.

    As he is going to remain in custody whilst witnesses are to be interviewed, the Custody Officer decides that Wayne is to be searched.

What do the Codes of Practice say about searching prisoners and the retention of their property?

....

....

....

►113

52►**26** It is important for the Custody Officer to inform the appropriate adult of his or her responsibilities as explained in Code C, para. 3.18 and you ensure that this has been done.

    The solicitor, Mr Edwards, wishes to speak to you before any interview.

What do you tell him?

....

....

....

What are the possible consequences of what you say?

....

....

....

►8

94▶ **27**

27

The evidence available is as follows:

- Identified by witness in an identification parade.

- Implicated by accomplice. (It is recognised that this does not strictly amount to evidence, or would not at a trial.)

- House searched, property from Sutherland Crescent burglary found and seized.

- Suspect's lock-up garage searched, property from Overton Gardens found and seized.

The Custody Officer looks at the evidence and statements and agrees that there is sufficient evidence to charge Peter Grey with both burglaries.

He is charged and bailed to Sandford Magistrates' Court. All you have to do now is complete the file.

THE END

109▶ **28**

28

What does PACE say about what you should do when you have evidence of further offences?

....

....

....

....

▶82

22▶ **29**

29

Having obtained the above information, the Custody Officer informs you that the solicitor and mother have arrived and are having a private consultation with Wayne.

CUSTODY RECORD	
Time	Full details of any action/occurrence involving detained person (Include full particulars of all visitors/officers) Individual entries need not be restricted to one line All entries to be signed by the writer (include rank and number)
18.50	Mother and solicitor arrived. Allowed private consultation with suspect. A. West PS
19.00	Concluded, remaining in interview room with above. A. West PS

▶52

3► 30 It is essential that you provide the Custody Officer with sufficient evidence to justify the arrest and that you have acted lawfully in how that arrest was made. 30

This does not have to be done in the presence of the suspect. You do not have to disclose any of the evidence to the suspect or their solicitor. To do so may have a detrimental effect on any subsequent interview.

Upon arrest the suspect must be informed of the grounds for the arrest. This does not extend to including any of the evidence. In this case, 'You have been arrested for burglary at 'Aysgarth', Overton Gardens, Sandford on 19th February' would be the grounds of arrest.

Having established that you have made a lawful arrest, what are the reasons when detention can be authorised by the Custody Officer?

....

....

....

....

►56

91► 31 The solicitor, Mr Edwards, (the Duty Solicitor), arrives at the station and is informed by the Custody Officer that he has already advised the co-defendant in this case. 31

What do the Codes of Practice say about solicitors advising co-defendants?

....

....

....

....

►114

It is now 2130 hrs and having got both offenders in custody, list what information and evidence you have available?

Offender 1—Wayne Cooper

....
....
....
....
....
....
....
....
....
....
....
....
....
....

Offender 2—Peter Grey

....
....
....
....
....
....
....
....
....
....
....
....
....

108▶ **33**   Section 17 PACE and Code B, paras 6.9 to 6.11 fully explain the extent to which any search can take place. This applies to all searches of premises.

Having entered the premises you find the youth hiding in a wardrobe in an upstairs bedroom. You arrest him on suspicion of burglary and caution him. He makes no reply.

What do you do now?

Take him straight back to the Police Station.   ▶125
Interview him at his home and then take him to the Police Station.   ▶88
Search the house.   ▶119

33

15▶ **34**

Name:	Wayne Cooper
Age:	16 years
Address:	14 Windsor Street, Sandford
Description:	5′ 8″ tall
	Fair collar-length hair.
	Slim build.
	Pierced left ear.
Vehicles:	None
Associates:	Graham Jones
	17 Miln Street, Sandford (Map Ref E6)
	21 years
	Previous for burglary, theft of cars, drugs.
	6′ 2″ tall, large build, black shoulder length hair, tattoos on knuckles of both hands.
	Brian David Thompson
	2 Hop Street, Sandford (Map Ref D6)
	18 years
	Previous for theft, robbery, public order.
	5′10″ tall, slim build, short fair hair, spiders web tattoo on right side of neck.
	Peter David Grey
	7 Patons Lane, Sandford (Map Ref D7)
	19 years
	Previous for burglary, handling, theft.
	5′11″ tall, slim build, very short fair hair.

▶22

34

129▶ **35**   Section 1 of PACE defines your powers of stop and search and where these powers may be exercised.

35

In this case, what are your reasonable grounds for suspicion?

....

....

....

....

▶73

1▶ **36**   By doing this, you allow the PC to return to her duties. However, you have then left yourself with all the scene of crime enquiries. It may be worth considering if the PC was in a position to help you.

36

▶50

81▶ **37**
- Description given by a witness.
- Stop and search—stolen property from burglary recovered.
- Interviewed in presence of mother and solicitor.
  – Admitted involvement in burglary at Overton Gdns.
  – Corroborated by knowledge of premises, property stolen and method of entry.
  – Informed police of the location of outstanding property. Confirmed by search.
  – Provided the name of his accomplice, Peter Grey. Detained and property from a burglary in Sutherland Crescent found during premises search.
- Re-interviewed re: involvement in the Sutherland Crescent burglary in presence of mother.
  – Admitted the offence.
  – Corroborated by knowledge of premises, property stolen and method of entry.
  – Confirmed involvement of Peter Grey.
- Statements from witness, Mrs Woodhams and victim, Mr Saunders.

37

What more do you require before you are in a position to ask the Custody Officer to consider charging Wayne?

....

....

....

....

....

▶54

112► **38**

Code A, para. 3.4 says that 'the search must be conducted at or nearby the place where the person or vehicle was first detained'.

In this case, the Police Station is too far away.

38

►112
make another choice

103► **39**

Description given by a witness.

Stop and search—stolen property from burglary recovered.

Interviewed in presence of mother and solicitor.

39

- Admitted involvement in burglary at Overton Gdns.

- Corroborated by knowledge of premises, property stolen and method of entry.

- Informed police of the location of outstanding property.
  – Confirmed by search.

- Provided the name of his accomplice, Peter Grey.
  – Detained and property from a burglary in Sutherland Crescent found during premises search.

Re-interviewed re: involvement in the Sutherland Crescent burglary in presence of mother.

- Admitted the offence.

- Corroborated by knowledge of premises, property stolen and method of entry.

- Confirmed involvement of Peter Grey.

Statements from witness, Mrs Woodhams and victim, Mr Saunders.

What more do you require before you are in a position to ask the Custody Officer to consider charging Wayne?

....

....

....

....

....

►123

WESTSHIRE CONSTABULARY

WITNESS STATEMENT

(C.J. Act, 1967, S.9; M.C. Act 1980,
S.102; M.C. Rules, 1980, r.70)

STATEMENT OF .... Sylvia Woodhams ....

AGE .... Over 18 years ....

This statement (consisting of . . . 2 . . . pages, each signed by me) is true to the best of my knowledge and belief and I make it knowing that if it is tendered in evidence, I shall be liable to prosecution if I have stated in it anything which I know to be false or do not believe to be true.

DATED .... 19th day of February ....

SIGNED .... Sylvia Woodhams ....

About 3.30 pm on Friday 19th February, I was sat at the desk in my office which is situated at the front of my home, 'Peppermill', Overton Gardens, Sandford. The office is on the first floor and overlooks my front garden and the houses opposite. I was on my own in the house.

As I was working at my desk, I looked out of the window and saw two male youths walking down the road from the direction of Elmwood Road towards the end of the cul-de-sac. They were about 20 yards away from my house but on the other side of the road.

The first youth I would describe as a white male, about 5'8" tall with fair collar length hair. He was wearing a green bomber type jacket. I would estimate him to be about 16 to 18 years old.

The second youth was also a white male, but slightly taller and older than the first. He was wearing blue jeans and a red rugby shirt. He had very short hair, almost a skinhead.

I didn't take much notice of them at first and I looked back at my desk and continued to work. I didn't see where they went or what they did.

About 3.40 pm the same afternoon, while I was still in my office, I again saw these two youths in the road outside. This time they were walking the other way towards Elmwood Road. The younger one was now carrying something in a white carrier bag. I can't remember if anything was written on the outside. I watched them as they walked down the road and out of my sight. They did not appear to be rushing and they seemed to be talking as they went.

Although I live in a quiet cul-de-sac, it is quite common to see door to door sales people and youths offering craft for sale. Nobody called at my house. I thought this was quite strange at the time, but was too busy with my work to give it a second thought.

From where I was sat in my office, the view into the road is clear and unobscured. The light was good and the distance between me and the youths was no more than 30 yards.

I have never seen the youths before. I would recognise them again if I saw them.

SIGNED .... Sylvia Woodhams ....

23► **41**

Having earlier considered searching Wayne's home and the reasons for the search you decide to speak to the Duty Inspector and explain the circumstances to her. She agrees with you and authorises the search to look for stolen property obtained as a result of other burglaries. She signs the required authorisation.

Wayne wants his mother to be present during the search.

What do the Codes of Practice say in relation to the presence of a nominated person during a search?

....

....

....

►70

18► **42**

What is the evidence that you have to put before the Custody Officer?

....

....

....

....

....

►80

41

42

113► 43

43

Code C, Annex A looks at intimate and strip searches. Subsection 'B' directly relates to strip searches.

Under the circumstances of this case, the Custody Officer authorises a strip search to establish that the subject does not have any more items of stolen property in his possession.

CUSTODY RECORD	
Time	Full details of any action/occurrence involving detained person (Include full particulars of all visitors/officers) Individual entries need not be restricted to one line All entries to be signed by the writer (include rank and number)
16.50	Strip search authorised by me in order to ascertain if the subject is in possession of other items of stolen jewellery. Search conducted, no further items found. A. West Ps.
16.55	Subject placed in detention room. A West Ps.

How might the conditions of detention affect your planning of any forthcoming interview?

....

....

....

....

....

►98

112► 44

44

You've been watching too many cop shows. You must have known that this was the wrong answer. Still, at least you're showing a sense of adventure in the way you work through this section.

►112

make another choice

14► **45** As is outlined in Code D, para. 3.2b, you should take care not to direct the witness's attention to any individual. Where there is sufficient information known to justify the arrest of a particular person for suspected involvement in the offence, that person is 'known' and Code D, para. 3.2 does not apply (Code D, para. 3.4).

►122

45

99► **46** Speak to the Duty Inspector and request authority to search the suspect's house.

46

What powers to you have to search the offender's house and what conditions have to be satisfied before you can do this?

....

....

....

....

....

►72

105► **47** Section 38 PACE, although explaining the Custody Officer's duties, it is important that you are aware of what is likely to happen to your prisoner.

47

What is the likely outcome with Wayne?

....

....

....

0020 hrs   Wayne is charged and bailed to appear at court in three weeks' time. His property is returned and he is released into the custody of his mother.

►18

91► **48** You only have circumstantial evidence to suggest that Wayne may have been involved in the other burglary.

    You interview him in the presence of his mother and he admits nothing. You return him to the detention room and consider searching his home in relation to outstanding property from Sutherland Crescent.

►66

48

Section 18(6) PACE outlines what you must inform an officer of the rank of Inspector or above, and Code B, paras 8.1 and 9.1 refer to the documentation that you are required to complete as a result of conducting a search.

You conduct the search and the outstanding property from the 'Aysgarth' burglary is found. You take your suspect to the Police Station and explain the circumstances of the arrest to the Custody Officer.

WESTSHIRE CONSTABULARY CUSTODY RECORD
Police and Criminal Evidence Act 1984

Police Station: SANDFORD

Force / Station Reference:

Station designated ? YES/NO

Reasons for Arrest: Suspicion of burglary AYSGARTH, OVERTON GDNS. SANDFORD ON 19/2.

Details of Detained Person

Surname: (Mr. Mrs. Miss.) GREY

First Name(s): PETER DAVID

PRISONER'S RIGHTS
A notice setting out my rights has been read to me and I have also been provided with a written notice setting out my entitlements while in custody.

Signature Peter Gray
Time 21·20   Date 19·2

Address: 7, PATONS LANE, SANDFORD, WESTSHIRE

Occupation: UNEMPLOYED

Appropriate adult / Interpreter   N/A

Time   Date

Age: 19 YRS   D.o.B 4·10·

Place of Birth: SANDFORD

Height: 5' 9"   Sex: M/F   M

Notification of named person

Requested [ ]   Not requested [✓]

Details of nominated person: N/A

Time   Date

Arrested by: Name:

Rank: DC   Sub-div./Dept SANDFORD

	Time	Date
Arrested at :	20·30	19/2
Arrived at station:	21·15	19/2

I want a solicitor as soon as practicable   ✓

Signature Peter Gray
Time 21·20   Date 19·2·

Appropriate adult / Interpreter   N/A

Time   Date

I do not want a solicitor at this time: N/A

Signature
Time   Date

Officer in the case: Name :

Rank/No./Sub-div./Dept : DC   SANDFORD.

Appropriate adult / Interpreter   N/A.

Time   Date

Officer opening Custody Record
Signature   a. West
Name   ALLAN WEST
Rank / No   PS 5603

Property
Bag No SF/91/9
All property to be listed.

Cash	Amount	
Notes	£	p
£50		
£20		
£10		
£5	5	00
Coins		42
Total	5	42

SEIZED	*RETAINED
(1) BROWN LEATHER WALLET CONTAINING VARIOUS PERSONAL PAPERS.	(1) ONE TIMEX WATCH (1) GOLD COLOURED METAL EARRING (1) ———"——— RING.

The above is a true record of my property *(items indicated have been retained by me) : Peter Gray

Signature of Search Officer : Peter Clayton PC 4901

Received the above property which is correct :

Witness (Police Officer) : a. West PS

Record action overleaf   *Delete as appropriate   In terrorism cases, all police officers' names should be replaced by their warrant number.

(continued)

49

	CUSTODY RECORD
Time	Full details of any action/occurrence involving detained person (Include full particulars of all visitors/officers) Individual entries need not be restricted to one line All entries to be signed by the writer (include rank and number)
21·20	Detention authorised by me in order to secure evidence from witness and obtain evidence by questioning. Subject states he is not suffering from any illness or injury. Dressed in red rugby shirt, blue jeans and a dark blue padded jacket. Delay between time of arrest and arrival at Police Stn. due to house and garage search conducted in his presence. A. West PS.

Peter does not want anyone informed of his detention. However, he would like the Duty Solicitor to attend. The Custody Officer obliges and makes the necessary phone call.

Now you are at the station, what should you do in relation to the conversation held between you and the suspect prior to arrival at the Police Station?

....

....

....

....

....

►76

1►50
36 By adopting this approach, you have two pairs of hands to conduct the enquiries. The PC can conclude obtaining the details for the crime report while you start house to house enquiries.

►132

118► **51**

Questioning suspects about the offence after they have been arrested and before arrival at the Police Station is not only bad practice, it is likely to render your evidence inadmissible. Code C, para. 11.1 explains the limitations of the questions that can be asked.

That covers the circumstances where you want to ask questions of the suspect, but what should you do if the suspect wants to talk to you about the offence for which he has been arrested?

....

....

....

....

►3

51

29► **52**

The Custody Officer informs you that the youth is ready for interview.

What should the Custody Officer advise the mother regarding her role as an appropriate adult?

....

....

....

....

....

►26

52

53

Time	Full details of any action/occurrence involving detained person (Include full particulars of all visitors/officers) Individual entries need not be restricted to one line All entries to be signed by the writer (include rank and number)
22.00	Custody to D.C........ for purpose of interview, in presence of his solicitor MR EDWARDS. A. West Ps.
22.15	Interview concluded PACE and Codes of Practice complied with. Subject returned to his cell.

CUSTODY RECORD

The interview takes place and Peter Grey chooses to make no comment.

How much of your interview preparation took account of this possibility? Where in the PEACE Model do you go to when a suspect indicates they are going to make no comment?

....

....

....

....

....

....

▶85

37▶ 54 | The Custody Officer listens to what you say, reads the statements and agrees to charge Wayne with both burglaries. | 54

What does PACE say about what must happen to a person after charge?

....
....
....
....
....

▶63

122▶ 55 | | 55

What evidence do you have to make an immediate arrest?

....
....
....
....
....
....

▶127

30▶ 56 | Although it is the Custody Officer's job to know these, it is also important for you to understand them. Section 37 PACE outlines the various reasons. You may feel that you want to interview a suspect to obtain evidence, but if the Custody Officer feels that there is sufficient evidence to charge, then detention must be authorised for the purpose of charging and not for obtaining evidence by questioning. | 56

Which reason would be applicable to the circumstances here, and where and when should these reasons be recorded?

....
....
....
....
....
....

▶74

9►
121 **57** Peter is bailed to return in two days' time at 6 pm. You are on from 2 pm to 10 pm which allows you time to do the final arrangements before he arrives.

57

For now, you still have to deal with Wayne Cooper. Would you:

Re-interview him about his possible involvement in the Sutherland Crescent burglary?

►23

Speak to the Duty Inspector with a view to searching his home for property from Sutherland Crescent?

►69

32► **58** ## Wayne COOPER—16 yrs

- Witness description.
- Stop and search—stolen property recovered.
- Arrested on suspicion of burglary at Overton Gdns.
- Detained at Police Station.
- Stop and search register completed.
- Notification delayed—authorised by Supt.
- Mother at Police Station.
- Duty Solicitor—Mr Edwards attended for interview.
- Interviewed in presence of mother and solicitor.
  - Admitted involvement in burglary at Overton Gdns.
  - Corroborated by knowledge of premises, property stolen and method of entry.
  - Location of outstanding property obtained. Property located.
  - Provided the name of his accomplice. Peter Grey.
- Remaining in custody until outcome of enquiries.
- Currently in custody 4 hours 45 minutes.

Peter David GREY—19 yrs

- Witness description.
- Implicated by accomplice.
- Arrested at home on suspicion of burglary at Overton Gdns.
- House searched, property found, believed stolen, seized.
  - Property from burglary—Sutherland Crescent.
- Suspect makes no comment during arrest and search.
- Suspect's lock-up garage searched before attending Police Station.
  - Property from 'Aysgarth' found and seized.
- Arrived at Police Station and detention authorised.
- Requested Duty Solicitor.
- Notes made of questions asked after arrest and during search.
- 'Premises searched register' completed.

DC Lee has obtained statements from Mr Saunders and Mr Attwood.

(continued)

58

What must be considered when deciding whether or not to return property to the loser?

....

....

....

....

►67

77►59

You may have a very commanding voice, but bearing in mind that the youth may not want to see you right now, is it realistic to expect him to come out to speak to you?

Why did you stop at the door? What powers of entry do you have under these circumstances?

....

....

....

....

►83

1►60

What impression of CID is the PC likely to have as a result of you checking up on her in that way?

....

....

....

Consider how you could involve the officer in your investigation rather than ignoring her?

....

....

....

►1
make another decision

79► **61** It is anticipated that you would carry out the following prior to an interview with Wayne:

- Obtain previous convictions.
- Go to local intelligence records.
- Obtain a statement from Mr Saunders including the identification of the item recovered.

►15

61

16► **62** Code C, para. 11.1A explains what constitutes an interview. Code C, para. 10.1 also outlines that you can ask questions which are confined to the proper and effective conduct of a search.

Peter refuses to comment on any of your questions, although he is present when the property is found.

When searching the suspect's coat, which is hanging on the back of his bedroom door, you find a set of keys, one of which has a small plastic tag with the number 5 on it. Again, Peter Grey says nothing.

You suspect that this is the key to the lock-up garage around the corner. You decide to take Peter to the Police Station via the garage.

62

What are your powers to search the garage?

....

....

....

....

....

....

►104

54► **63** Section 38 PACE, although explaining the Custody Officer's duties, it is important that you are aware of what is likely to happen to your prisoner.

63

What is the likely outcome with Wayne?

....

....

....

0020 hrs   Wayne is charged and bailed to appear at court. His property is returned and he is released into the custody of his mother.

►5

$^{96}_{99}$▶ **64** Check the location of the outstanding property.

2010 hrs You and a colleague attend the block of lock up garages in Thompson Street. You speak to a middle aged man who is parking his car in garage No. 4. You ask him if Peter Grey still rents garage No. 5. He replies, 'If that's the lad from Patons Lane, then yes, but I've not seen him use it for a while'.

| 64

What are your powers to search the garage at this time?

....

....

....

....

....

....

▶20

101▶ **65** Code C, para. 10.1 (c) outlines that questioning which is confined to the proper and effective conduct of a search is not improper.

| 65

Under what power can you seize and retain property found as a result of the search?

....

....

....

....

▶21

$^{48}_{91}$▶ **66** Having earlier considered searching Wayne's home and the reasons for the search, you decide to speak to the Duty Inspector and explain the circumstances to her. She agrees with you and authorises the search to look for stolen property obtained as a result of similar offences. She signs the required authorisation.

Wayne wants his mother to be present during the search.

| 66

What do the Codes of Practice say in relation to the presence of a nominated person during a search?

....

....

....

....

....

▶75

⁵⁸► **67**   Section 22 PACE and Code B, paras 7.14 and 7.15 explain your position in relation to the retention and disposal of property subject of crime.

67

►4

¹⁸► **68**

68

What would be the purpose of the interview?

....

....

....

....

►94

⁵⁷► **69**   Having earlier considered searching Wayne's home and the reasons for the search, you decide to speak to the Duty Inspector and explain the circumstances to her. She agrees with you and authorises the search to look for stolen property obtained as a result of other burglaries. She signs the required authorisation.

69

Wayne wants his mother to be present during the search.

What do the Codes of Practice say in relation to the presence of a nominated person during a search?

....

....

....

►95

41► **70**   Code B, para. 5.11.   **70**

Under these circumstances, what options do you have?

....

....

....

How does s. 18 PACE affect which rooms of the house you can search?

....

....

....

This may actually be affected by the relationship between the juvenile and his parents as to whether or not their bedroom can be searched.

Mrs Cooper accompanies you to the house and is present while you search Wayne's bedroom. You don't find anything from Overton Gardens. However, in the top drawer of his bedside cabinet, you find a cheque card from the burglary in Sutherland Crescent.

You return to the Police Station with Mrs Cooper with a view to arresting Wayne for the burglary and interviewing him.

►109

77► **71**   **71**

What are your powers of entry?

....

....

....

►83

46► **72** | Section 18 PACE outlines the powers and Code B para. 4.3 the conditions attached. | **72**

How do the circumstances of this offence fit to these conditions?

....
....
....
....
....
....

What would you say to the Duty Inspector to justify obtaining the required authority?

....
....
....

►96

35► **73** | In these circumstances, the proximity of the offence, the similarity of the description and seeing him hide something in his pocket as he saw you. | **73**

What must you inform the youth of before conducting the search?

....
....
....

►11

WESTSHIRE CONSTABULARY CUSTODY RECORD
Police and Criminal Evidence Act 1984

Police Station: SANDFORD	**Force / Station Reference:**	**Station designated ?** YES/~~NO~~

Reasons for Arrest: BURGLARY
AXBGAETH OVERTON CONS, SANDFORD

PRISONER'S RIGHTS
A notice setting out my rights has been read to me and I have also been provided with a written notice setting out my entitlements while in custody.

Signature ... Wayne Cooper
Time ... 16·43 ... Date ... 19·2·

Appropriate adult / Interpreter

Time .... Date ....

Notification of named person

Requested ☑   Not requested ☐

Details of nominated person: GRAHAM JONES

Time 16·43   Date 19·2·

I want a solicitor as soon as practicable

Signature Wayne Cooper
Time .... Date ....

Appropriate adult / Interpreter

Time .... Date ....

I do not want a solicitor at this time:

Signature ....
Time .... Date ....

Appropriate adult / Interpreter

Time.................... Date ....

Officer opening Custody Record
Signature   A. West
Name   BRIAN WEST
Rank / No   PS 5603

Details of Detained Person

Surname: (Mr. ~~Mrs. Miss~~.) Cooper

First Name(s): WAYNE

Address: 14, WINDSOR ST
SANDFORD,
WESTSHIRE

Occupation: UNEMPLOYED

Age: 16   D.o.B 23/11/

Place of Birth: SANDFORD

Height: 5'3"   Sex: M/~~F~~   M

Arrested by: Name: ....

Rank: DC   Sub-div./Dept SANDFORD

	Time	Date
Arrested at:	16·25	19/2
Arrived at station:	16·40	19/2

Officer in the case: Name: ....

Rank/No./Sub-div./Dept: DC
SANDFORD

CUSTODY RECORD

Time	Full details of any action/occurrence involving detained person (Include full particulars of all visitors/officers) Individual entries need not be restricted to one line All entries to be signed by the writer (include rank and number)
16·45	Detention authorised by me in order to secure evidence from witnesses and obtain evidence by questioning. Subject states he is not suffering from any illness or injury. Dressed in green bomber type jacket, navy blue shell suit bottom and a white 'T' shirt. A. West PS.

(continued) 74

What legal rights does the suspect have when in police detention?

....
....
....
....
....

►7

66► 75

Code B, para. 5.11.

Under these circumstances, what options do you have?

....
....
....

How does s. 18 PACE affect which rooms of the house you can search?

....
....
....

This may actually be affected by the relationship between the juvenile and his parents as to whether or not their bedroom can be searched.

Mrs Cooper accompanies you to the house and is present while you search Wayne's bedroom. In the top drawer of his bedside cabinet, you find a cheque card from the burglary in Sutherland Crescent.

You return to the Police Station with Mrs Cooper with a view to arresting Wayne for the burglary and interviewing him.

►116

75

49▶76 Code C, para. 11.13 states that you should make a written record of any comments made by a suspected person, including unsolicited comments, which are outside the context of an interview, but which might be relevant to the offence. It goes on to say that the record should be timed and signed by the maker and where practicable that the person should be given the opportunity to read the record and to sign it as correct or to indicate the respects in which he considers it inaccurate. Code C, Note 11E states the suspect should be asked to endorse the same with the words such as 'I agree that this is a correct record of what was said'.

In the situation that you have, Peter has declined to comment. It would be advisable to write up notes of the questions that you asked and when you asked them. These notes could then be used in any subsequent interview, where they could be offered to him to comment on, in accordance with Code C, para. 11.24.

As a result of enquiries from recent crime reports, you establish that the jewellery and credit cards found in Peter's house was from a burglary yesterday in Sutherland Crescent, Sandford (Map Ref B4).

WESTSHIRE CONSTABULARY

CRIME REPORT

Crime Reference

Rec Stn Ref	Number	Year	Suff

Division: Sub-Division: Beat:

1. When Reported	Time (24hr)	Day	Date	Month	Year	2. Offence as Reported	Att	S. HO Class
	2000	THUS	18	2		BURGLARY (DWELLING)		

4. Name of Person Reporting: DAVID ATTWOOD

5. Address of Person Reporting: 17, SUTHERLAND CRESCENT, SANDFORD

6. How Report Received (Tick ✓ appropriate box)

Aggrieved Person			Other Person		
Police Station	1	✗	Police Station	5	□
"999" Call	2	□	"999" Call	6	□
Other Telephone	3	□	Other Telephone	7	□
To Police Patrol	4	□	To Police Patrol	8	□
Other Means			Burglar Alarm		
On Admission	9	□	Audible	13	□
Other Inquiries	10	□	"999"	14	□
Found by Police	11	□	Direct	15	□
Observation	12	□	Actuated	Yes 16	□
				No 17	□

Postal Code   **Telephone Number** 447858

7. Officers Involved

	Rank	Number	Names
To whom reported	PC	6404	CARTER
Investigating	DC		
	SOCO	AMANDA	RICHARDSON

AGGRIEVED PERSON

8. Name of Aggrieved Person	Age	Sex
DAVID ATTWOOD	65	M

9. Occupation of Aggrieved Person RETIRED TEACHER

10. Address of Aggrieved Person 17, SUTHERLAND CRESCENT SANDFORD

11. Name of Aggrieved Company/Business

12. Type of Aggrieved Company/Business

13. Address of Aggrieved Company/Business (or business address of aggrieved person)

Postal Code   **Telephone Number** 447858   **Postal Code**   **Telephone Number**

PLACE OF OFFENCE

14. Address of Offence AS ABOVE

15. Location/Type of Premises DETACHED DWELLING HOUSE

16. When Comm.	Time	Day	Date	Month	Year
At/Between	1300	THUS	18	2	
and	2000	– " –	– " –	– " –	

DETAILS OF OFFENCE

17. Modus Operandi Offender(s) went to rear of detached house in quiet cul-de-sac. Smashed small ground floor window, opened larger window, gained entry. Untidy search of all rooms, exit as entry. (Front and rear doors locked). Property stolen.

18. Types of Weapon Used N/A

19. Injury to Victim Fatal 1 □ Serious 2 □ Slight 3 □ Threats 4 □ None 5 ✗

20. Special details (tick as necessary) N/A. Drugs □ Firearms □ 1 □ 2 □ 3 □ 4 □ 5 □

21. PROPERTY STOLEN/DAMAGED

Description		Value		
		Stolen	Recovered	Damaged
'SEE PROPERTY CONTINUATION SHEET'		£300	NIL	£30
Continue on "Stolen Property Continuation Sheet" if necessary	TOTAL	£300	NIL	£30.

Investigation by	Uniform 01 □	Crime			
	CID 02 ✗	Screening	In 01 ✗	Out 02 □	Not Applicable 03 □

(continued) 76

Stolen Property Continuation Sheet		
Identifiable	Non-identifiable	Recovered
Access Credit Card Nº 200947732235 Nat West Cheque Card Nº 3476436X9 Yellow Gold w/Band Engraved. DA 14/11/52 (Value £250)	Assorted Costume Jewellery (Value £50)	

▶32

20▶ 77

Without a warrant, you would need to have arrested the person who controls the garage; in this case, Peter Grey. Then you have to use your powers under s. 18 PACE.

As Peter only lives around the corner, it makes sense to see if he is in.

You then drive around the corner to 7 Patons Lane. As you pull up outside you see a male youth walking down the front path towards the road. He is in his late teens or early twenties, with a skinhead haircut. He's wearing blue jeans and a red rugby shirt. On seeing you, the youth turns and runs back into the house leaving the front door ajar. What do you do now?

Let the youth go back into the house, shrug your shoulders and leave. ▶133

Shout at the youth to stop and wait at the door for an answer. ▶59

Run after the youth and search the house. ▶71

WESTSHIRE CONSTABULARY

CRIME REPORT

	Crime Reference			
	Rec Stn Ref	Number	Year	Suff

Division: Sub-Division: Beat:

1. When Reported	Time (24hr)	Day	Date	Month	Year
	1545	FRI	19	2	

2. Offence as Reported	Att	S. HO Class
BURGLARY (DWELLING)		

4. Name of Person Reporting:
MARTIN SAUNDERS

5. Address of Person Reporting:
'AYSGARTH'
OVERTON GARDENS
SANDFORD

Postal Code	Telephone Number
	447852

7. Officers Involved

	Rank	Number	Names
To whom reported	PC	4808	HOPKINS
Investigating	DC		
	SOCO	AMANDA	RICHARDSON

6. How Report Received (Tick ✓ appropriate box)

Aggrieved Person			Other Person		
Police Station	1	☒	Police Station	5	☐
"999" Call	2	☐	"999" Call	6	☐
Other Telephone	3	☐	Other Telephone	7	☐
To Police Patrol	4	☐	To Police Patrol	8	☐
Other Means			Burglar Alarm		
On Admission	9	☐	Audible	13	☐
Other Inquiries	10	☐	"999"	14	☐
Found by Police	11	☐	Direct	15	☐
Observation	12	☐	Actuated	Yes 16	☐
				No 17	☐

AGGRIEVED PERSON

8. Name of Aggrieved Person	Age	Sex
MARTIN SAUNDERS	47 yrs	M

9. Occupation of Aggrieved Person
COMPUTER ANALYST

10. Address of Aggrieved Person
'AYSGARTH'
OVERTON GARDENS
SANDFORD

Postal Code	Telephone Number 447852

11. Name of Aggrieved Company/Business

12. Type of Aggrieved Company/Business

13. Address of Aggrieved Company/Business
(or business address of aggrieved person)

Postal Code	Telephone Number

PLACE OF OFFENCE

14. Address of Offence
'AYSGARTH'
OVERTON GARDENS
SANDFORD

15. Location/Type of Premises
DWELLING HOUSE (DETACHED)

16. When Comm.	Time	Day	Date	Month	Year
At/Between	15.15	FRI	19	2	
and	15.45	"	"	"	"

DETAILS OF OFFENCE

17. Modus Operandi Offender(s) went to rear of detached house situated in cul-de-sac. Smashed small ground floor window then opened larger window, gained entry. Very untidy search of all rooms, property stolen. Exit as entry. Front and back doors were locked.

18. Types of Weapon Used N/A

19. Injury to Victim
Fatal 1 ☐ Serious 2 ☐ Slight 3 ☐ Threats 4 ☐ None 5 ☒

20. Special details (tick as necessary) N/A.
Drugs ☐ Firearms ☐ 1 ☐ 2 ☐ 3 ☐ 4 ☐ 5 ☐

21. PROPERTY STOLEN/DAMAGED

Description		Value	
	Stolen	Recovered	Damaged
'SEE CONTINUATION SHEET'	£5,400	NIL	£30
Continue on "Stolen Property Continuation Sheet" if necessary TOTAL	£5,400	NIL	£30

Investigation by	Uniform 01 ☐	Crime			
	CID 02 ☒	Screening	In 01 ☒	Out 02 ☐	Not Applicable 03 ☐

(continued) 78 | 78

Stolen Property Continuation Sheet		
Identifiable	Non-identifiable	Recovered
HITACHI MINI HIFI COMPACT DISC PLAYER & SPEAKERS Nº/ H/S129/8793·3645 £400	WHITE GOLD ENGAGEMENT RING SET WITH 8 SAPPHIRES & 8 DIAMONDS	
ACCESS CARD MR. M. SAUNDERS Nº 2775 0078 9244 7408	18CT GOLD CHAIN WITH 18CT GOLD FOB PENDANT	
VISA CARD MR. M. SAUNDERS Nº 2996 2775 8264 0024	9CT GOLD FIVE BAR GATE BRACELET	
NAT WEST CHEQUE BOOK M & P SAUNDERS A/C Nº 47735216 SORT CODE :– 12·22·06	18CT GOLD RING SET WITH A SINGLE DIAMOND 18CT GOLD LADIES ROLEX WRIST WATCH TOTAL VALUE £5000-00p.	

►100

90► 79 Code C, para. 6.6 and Annex B outlines under what circumstances you can conduct an urgent interview. The Superintendent is not convinced and he declines your request. | 79

You have until about 6.30 pm to prepare for the interview with Wayne. Apart from looking at the areas to be addressed in the interview itself, what else can you be doing?

....

....

....

....

....

....

►61

42► 80 The evidence available is as follows: | 80

- Identified by witness in an identification parade.

- Implicated by accomplice.

- House searched, property from Sutherland Crescent burglary found and seized.

- Suspect's lock-up garage searched, property from Overton Gardens found and seized.

The Custody Officer looks at the evidence and statements and agrees that there is sufficient evidence to charge Peter Grey with both burglaries.

He is charged and bailed to Sandford Magistrates' Court. All you have to do now is complete the file.

THE END

13►**81** Section 31 PACE explains about the need to arrest the person if it appears to you that he would be liable to arrest for another offence if he was released.   **81**

Time	Full details of any action/occurrence involving detained person (Include full particulars of all visitors/officers) Individual entries need not be restricted to one line All entries to be signed by the writer (include rank and number)
CUSTODY RECORD	
23·10	Custody to D.C........ for the purpose of interview in the presence of his mother. A. West Ps
23·30	Interview concluded Subject returned to Detention Room PACE. and Codes of Practice complied with.

What evidence do you have to put before the Custody Officer for a decision to charge Wayne?

....

....

....

....

....

►37

28►82

Section 31 PACE explains about the need to arrest the person if it appears to you that he would be liable to arrest for another offence if he was released.

82

CUSTODY RECORD

Time	Full details of any action/occurrence involving detained person (Include full particulars of all visitors/officers) Individual entries need not be restricted to one line All entries to be signed by the writer (include rank and number)
23·10	Custody to D.C....... for the purpose of interview in the presence of his mother. A. West Ps.

2330 hrs   The interview was concluded. You return Wayne to the detention room and make the necessary entry on the custody record, not forgetting to include the fact that he has been arrested for the second burglary at Sutherland Crescent.

With the new evidence that you have put before him, Wayne fully admits the offence at Sutherland Crescent and can provide you with the details of the offence that only the person responsible would know. He also stated that the offence was committed with Peter Grey.

What evidence do you have to put before the Custody Officer for a decision to charge Wayne?

....

....

....

....

....

►115

⁵⁹⁄₇₁ ▶ **83**

Section 17(1)(b) PACE outlines your powers of entry in these circumstances.

83

What are the conditions under which you can exercise this power?

....
....
....
....
....

▶10

¹²² ▶ **84**

84

What are your powers of arrest under s. 24 PACE?

....
....
....
....
....
....

▶117

⁵³ ▶ **85**

You need to go straight to your prepared agenda of the PEACE Model. Peter Grey disputes the identification evidence and wants an identification parade.

85

What do the Codes of Practice say about his request for an identification parade?

....
....
....
....
....

▶121

¹¹⁶▶86

What does PACE say about what you should do when you have evidence of further offences?

....
....
....
....

▶103

¹²⁴▶87

Mr Edwards says that his client disputes the identification evidence.

What do the Codes of Practice say about where there is a dispute over identification?

....
....
....
....
....

▶92

³³▶88

What do the Codes of Practice say about when and where you can conduct an interview?

....
....
....
....
....

▶130

112▶89 | The Codes of Practice say that 'the search must be conducted at or nearby to the place where the person or vehicle was first detained'. (Code A, para. 3.4).

Provided the disused shop front is nearby, then there is no problem. |

What do the Codes of Practice say about the extent of the search and how does this relate to the circumstances we have here?

....

....

....

....

....

....

....

....

....

▶97

12▶90 Under the circumstances as explained, the Inspector agrees to a delay in notifying arrest.

▶90

CUSTODY RECORD	
Time	Full details of any action/occurrence involving detained person (Include full particulars of all visitors/officers) Individual entries need not be restricted to one line All entries to be signed by the writer (include rank and number)
17.30	Delay in notification authorised by me as this is a serious arrestable offence. (Sec 116 P.A.C.E.) in that there would be a substantial financial gain to the offenders and as there is still an outstanding offender, notification could result in him being alerted. This could also hinder the recovery of property obtained from this offence. A.O.Lee (Sergt)

While waiting for the juvenile's mother, a colleague tells you that you have a right to interview the suspect before the mother and solicitor arrive, because there is an offender still at large and stolen property still unrecovered.

Is this right?

....

....

....

▶79

4▶91 What would be your next courses of action?

▶91

Interview Peter Grey.

▶31

Re-interview Wayne Cooper regarding his involvement in the Sutherland Crescent burglary.

▶48

Speak to the Duty Inspector with a view to searching his home for property from Sutherland Crescent.

▶66

87►92 Code D, paras 3.12 and 3.14 says that where there is a dispute about identification an identification procedure must be held. It must be held as soon as practicable.

It is not practical to consider holding an identification procedure at this time of night. What are your options re: identification procedure?

Keep in custody until the morning and arrange the identification procedure.

►6

Bail to come back to the Police Station at a time when you will have had chance to arrange the identification procedure.

►19

92

123►93 Section 38 PACE, although explaining the Custody Officer's duties, it is important that you are aware of what is likely to happen to your prisoner.

93

What is the likely outcome with Wayne?

....

....

....

....

0020 hrs   Wayne is charged and bailed to appear at court in three weeks' time. His property is returned and he is released into the custody of his mother.

►131

68►94 The Codes of Practice, Code C, para. 16.1, says that once you reasonably believe that there is sufficient evidence to provide a realistic prospect of a conviction, AND you are satisfied that all questions you consider relevant to obtaining accurate and reliable information about the offence have been put, then taking account of any other available evidence, the interview must cease (11.6).

In these circumstances, you may wish to give Peter a final opportunity to say something.

94

He declines to comment. There is no purpose in the interview continuing. What evidence have you to present to the Custody Officer?

....

....

....

....

....

►27

69►95 | Code B, Para. 6.11. | 95

Under these circumstances, what options do you have?

....

....

How does s. 32 PACE affect which rooms of the house you can search?

....

....

....

This may actually be affected by the relationship between the juvenile and his parents as to whether or not their bedroom can be searched.

Mrs Cooper accompanies you to the house and is present while you search Wayne's bedroom. You don't find anything from Overton Gardens. However, in the top drawer of his bedside cabinet, you find a cheque book from the burglary in Sutherland Crescent.

You return to the Police Station with Mrs Cooper with a view to arresting Wayne for the burglary and interviewing him.

►106

72►96 | The offence for which Wayne has been arrested is an arrestable offence, there is property still outstanding from that offence, and although the offender has indicated where some of the outstanding property could be, it is suspected that some could be located at his home address. | 96

You do not have any firm evidence to implicate his involvement in any other arrestable offences connected to or similar to the burglary for which he is custody.

Although a search of Wayne's house would be an essential part of this investigation, is it the most logical next step when we have been told of the location of outstanding property and the name of his accomplice?

....

....

....

....

►64

Code A, paras 3.1, 3.3, 3.5, 3.6, 3.7 and Note 5 indicate the extent to which you can search.

In the circumstances of this incident, you have grounds to believe that he has hidden something in his trouser pocket. However, there are other grounds to suspect his involvement, such as proximity and description. Code A, para. 3.3 allows you to conduct an extensive search.

When you start to conduct the search, the youth resists by struggling. What do the Codes of Practice say about the use of force to conduct a search?

....

....

....

....

►101

Code C, para. 8 explains the basic conditions of detention and you should be aware of them when planning your investigation.

The Custody Officer then tries to contact the youth's parents and the Duty Solicitor.

CUSTODY RECORD

Time	Full details of any action/occurrence involving detained person (Include full particulars of all visitors/officers) Individual entries need not be restricted to one line All entries to be signed by the writer (include rank and number)
17.10	Mother informed and will attend in about 45 minutes. A. West PS.
17.15	Duty solicitor, Mr EDWARD (WaBurn) informed and will attend at 18.30 hrs. A. West PS

 As the officer in the case, what observations would you have on the request to have his friend informed of his arrest?

....

....

....

 What options are open to you if you want to challenge this request?

....

....

....

►12

$\frac{8}{40}$►99   It is important to recognise that admissions can be rendered inadmissible if they are obtained in an improper manner. Sections 76, 77 and 78 PACE outline the law relating to confessions and the exclusion of unfair evidence.

What are you going to do now?

Charge immediately.	►2
Obtain a witness statement from Mrs Woodhams.	►107
Conduct a search of Wayne's home.	►46
Check the location of the outstanding property.	►64

100

78►100   After getting no reply from the houses immediately next door, you go across the road to a house called 'Peppermill'. There, you speak to Mrs Sylvia Woodhams. Mrs Woodhams is a college lecturer who works from home. She has been in all afternoon and on being asked if she had seen anything out of the ordinary she recalls:

'About 3.30 pm I was sat at the desk in my office which is the upstairs front room overlooking my front garden and the houses opposite. I looked out of the window and saw two young lads walking down the road. They were on the other side of the road.

One was about 16 to 18 years old, 5'8" tall with fair collar length hair. He was wearing a green bomber type jacket. The other was slightly taller and older. He was wearing blue jeans and a red rugby shirt. He had very short hair, almost a skinhead. I didn't see where they went or what they did. About 10 minutes later I again saw these lads in the road outside. This time they were walking the other way, out of the cul-de-sac. The younger one was now carrying something in a white carrier bag. I watched them as they walked down the road and out of my sight. They were just walking and they seemed to be talking as they went. To be honest, I didn't give it another thought until now. I suppose I might recognise them again if I saw them; I don't know.'

Mrs Woodhams is willing to accompany you in your police car to check the local area to see if you can locate the offenders.

What rules do the Codes of Practice state must be adhered to in relation to this form of identification? What does the Code of Practice say about the first description?

....

....

....

....

....

....

....

....

....

....

►14

97►101

Code A, para. 3.2 outlines the use of reasonable force as a last resort.
  After a brief struggle, the youth calms down and you conduct the search.

1C

During the search, you speak to the youth. What restrictions are there on what you can ask of him?

....

....

....

....

....

►65

7►102

Code C, paras 3.12 to 3.20 add additional rights for juveniles and special groups of detained persons.

1C

Who may act as an appropriate adult?

....

....

....

....

....

►25

Section 31 PACE explains about the need to arrest the person if it appears to you that he would be liable to arrest for another offence if was released.

CUSTODY RECORD	
Time	Full details of any action/occurrence involving detained person (Include full particulars of all visitors/officers) Individual entries need not be restricted to one line All entries to be signed by the writer (include rank and number)
22.40	Custody to D.C...... for the purpose of interview in the presence of his mother. A. West PS.
23.00	Interview concluded. Subject arrested for an offence of burglary at SUTHERLAND CRES. Subject returned to Detention Room. P.A.C.E. and Codes of Practice complied with.

With the new evidence that you have put before him, Wayne fully admits the offence at Sutherland Crescent and can provide you with the details of the offence that only the person responsible would know. He also states that the offence was committed with Peter Grey.

What evidence do you have to put before the Custody Officer for a decision to charge Wayne?

....
....
....
....
....
....

▶39

62▶**104**

Section 18(5) PACE outlines your powers, but remember that Code B still applies.

Having exercised your powers, what must you do after completing such a search?

....
....
....
....
....
....
....
....
....
....

▶49

115▶**105**

What does PACE say about what must happen to a person after charge?

....
....
....
....
....

▶47

95►**106** While all this was happening, Wayne Cooper has been in custody for $5\frac{1}{2}$ hours. The Custody **106**
Officer needs to be updated with the state of the investigation.

Assuming that you would not want to release Wayne at the moment, what would you
have to explain to the Custody Officer?

The Custody Officer is happy that you are conducting this investigation expeditiously
and informs Inspector Hazel Atkinson that the review is due and tells her the circumstances
of the case.

CUSTODY RECORD	
Time	Full details of any action/occurrence involving detained person (Include full particulars of all visitors/officers) Individual entries need not be restricted to one line All entries to be signed by the writer (include rank and number)
22:30	Review of detention conducted by me. Further detention authorised for the purpose of obtaining evidence by questioning. I have informed the suspect of his continued right to free legal advice in the presence of his mother. The subject is prepared to be interviewed further without his solicitor being present. Insp Hazel Atkinson

What are the areas that you intend to question Wayne about during the second interview?

....

....

....

....

....

►13

99►**107** Obtain a witness statement from Mrs Woodhams. Because she feels part of this investigation **107**
due to the way you treated her earlier, PC Janet Hopkins offers to get this for you. This allows
you to complete any other enquiries that you feel are necessary.

►40

10▶**108**

Both s. 3 Criminal Law Act 1967 and s. 117 PACE explain the use of reasonable force.

To what extent can you search the premises?

....

....

....

....

....

▶33

While everything else is going on, Wayne Cooper has been in custody for $5\frac{1}{2}$ hours. The Custody Officer needs to be updated with the state of the investigation.

Assuming that you would not want to release Wayne at the moment, what would you have to explain to the Custody Officer?

The Custody Officer is happy that you are conducting this investigation expeditiously and informs Inspector Hazel Atkinson that the review is due and tells her the circumstances of the case.

CUSTODY RECORD	
Time	Full details of any action/occurrence involving detained person (Include full particulars of all visitors/officers) Individual entries need not be restricted to one line All entries to be signed by the writer (include rank and number)
22:30	Review of detention conducted by me. Further detention authorised for the purpose of obtaining evidence by questioning I have informed the suspect of his continued right to free legal advice in the presence of his mother. The subject is prepared to be interviewed further without his solicitor being present. Insp Hazel Atkinson

What are the areas that you intend to question Wayne about during the second interview?

....

....

....

....

▶28

17►**110** Code C, Note 6G explains that it is the solicitor who makes the decision about any possible conflicts of interest. As a general rule, you cannot make any observations unless Code C, Annex B and/or Code C, para. 6.6(b) applies.

In this case, Mr Edwards feels that there would not be any conflict of interests and the Custody Officer allows him a private consultation with Peter Grey.

What are the main areas that you are going to ask questions about?

....

....

....

....

....

....

►124

128►**111** Section 32(3), (4), (5) PACE and Code B, paras 6.9 to 6.11 cover the conditions under which you can conduct a search, while s. 32 PACE and Code B, para. 6.9B look at the limitations and conduct of the search.

You search the house and while looking in the upstairs front bedroom, you find a quantity of jewellery and credit cards in the name of David Attwood in a sock in a chest of drawers. You suspect this to be stolen, although not from the current burglary that you are investigating.

What are your powers to seize such property?

....

....

....

►16

11►**112** Having established that you are going to conduct the search, which of the following options would you take?

Take him to the Police Station for the purpose of the search.	►38
Take him to a nearby disused shop front and search all his outer clothing.	►89
Search the pocket in which you believe you saw him place an article.	►97
Automatically spreadeagle him against the nearest wall and search all his pockets.	►44

25►**113**   Code C, paras 4.1 to 4.3 explains.

Under what circumstances can more than a clothing search be made?

....

....

....

....

....

►43

31►**114**   Code C, Note 6G explains that it is the Solicitor who makes the decision about any possible
conflicts of interest. As a general rule, you cannot make any observations unless Code C,
Annex B and/or Code C, para. 6.6(b) applies.

In this case, Mr Edwards feels that there would not be any conflict of interests and the
Custody Officer allows him a private consultation with Peter Grey.

What are the main areas that you are going to ask questions about?

....

....

....

....

....

....

►53

82►115

- Description given by a witness.
- Stop and search—stolen property from burglary recovered.
- Interviewed in presence of mother and solicitor.
  - Admitted involvement in burglary at Overton Gdns.
  - Corroborated by knowledge of premises, property stolen and method of entry.
  - Informed police of the location of outstanding property. Confirmed by search.
  - Provided the name of his accomplice, Peter Grey. Detained and property from a burglary in Sutherland Crescent found during premises search.

- Re-interviewed re: involvement in the Sutherland Crescent burglary in presence of mother.
  - Admitted the offence.
  - Corroborated by knowledge of premises, property stolen and method of entry.
  - Confirmed involvement of Peter Grey.

- Statements from witness, Mrs Woodhams and victim, Mr Saunders.

What more do you require before you are in a position to ask the Custody Officer to consider charging Wayne?

....

....

....

The Custody Officer listens to what you say, reads the statements and agrees to charge Wayne with both burglaries.

►105

Wayne Cooper has been in custody for $5\frac{1}{2}$ hours. The Custody Officer needs to be updated with the state of the investigation.

Assuming that you would not want to release Wayne at the moment, what would you have to explain to the Custody Officer?

....

....

....

....

....

The Custody Officer is happy that you are conducting this investigation expeditiously and informs Inspector Hazel Atkinson that the review is due and tells her the circumstances of the case.

CUSTODY RECORD	
Time	Full details of any action/occurrence involving detained person (Include full particulars of all visitors/officers) Individual entries need not be restricted to one line All entries to be signed by the writer (include rank and number)
22:30	Review of detention conducted by me. Further detention authorised for the purpose of obtaining evidence by questioning. I have informed the suspect of his continued right to free legal advice in the presence of his mother. The subject is prepared to be interviewed further without his solicitor being present. Insp Hazel Atkinson

What are the areas that you intend to question Wayne about during the second interview?

....

....

....

....

....

....

▶86

84►**117**

Which of them is applicable to these circumstances?

....

....

....

....

....

►126

21►**118**

Code A, para. 4 explains the procedure required for the making of written records of searches made.

As a result of the search, you find a gold ring with a single diamond setting in his front left trouser pocket. The arrest is made and the suspect cautioned. Having arrested the youth, Wayne Cooper, you call for assistance to transport your prisoner to the Police Station. While waiting with him you have a conversation.

What are the limitations on the questions you can now ask?

....

....

....

....

►51

33►**119**

As you have Peter Grey with you and you will want to conduct a search of his house at some point in the investigation, it seems logical to conduct the search now. What powers do you have to carry out the search?

....

....

....

....

....

►128

24►**120**   Code A, para. 1.5 says that you cannot search a person, even with their consent, where no power to search is applicable.

►129

120

85►**121**   Code D, paras 3.12 and 3.14 says that where there is a dispute over identification an identification procedure must be held. It must be held as soon as practicable.

It is not practical to consider holding an identification procedure at this time of night. What are your options re: identification procedure?

Keep in custody until the morning and arrange the identification procedure?   ►9

Bail to come back to the Police Station at a time when you will have had chance to arrange the identification procedure?   ►57

121

45►**122**   Unfortunately, Mrs Woodhams has been unable to identify anyone and you have returned her to her house. On your way back to the station, you are driving along South Road (Map Ref B4–C4) when you see a 16 to 18 year old youth with collar length fair hair. He is wearing a green bomber type jacket. As you approach him, you see him place something in his front left trouser pocket.

Do you:

Stop him and immediately arrest him on suspicion of burglary?   ►55

Stop him, ask him where he has just been, what he has been doing and if not satisfied with his answers, arrest him?   ►84

Stop him, ask him where he has just been, what he has been doing and ask him if he would allow you to search him?   ►24

Stop him for the purpose of conducting a s. 1 PACE stop and search?   ►129

122

³⁹►**123**

The Custody Officer listens to what you say, reads the statements and agrees to charge Wayne with both burglaries.

What does PACE say about what must happen to a person after charge?

....

....

....

....

....

►93

¹¹⁰►**124**

CUSTODY RECORD	
Time	Full details of any action/occurrence involving detained person (Include full particulars of all visitors/officers) Individual entries need not be restricted to one line All entries to be signed by the writer (include rank and number)
22.00	Custody to D.C...... for the purpose of interview in the presence of MR EDWARDS (SOLICITOR) A. West PS
22.15	Interview concluded. P.A.C.E and Codes of Practice complied with. Subject returned to his cell.

The interview takes place and Peter Grey chooses to make no comment.

How much of your interview preparation took account of this possibility? You may find it useful to refer to the 'Preparation and Planning' section of the 'A Practical Guide to Interviewing' handbook.

....

....

....

....

►87

³³▶**125** | You get him back to the Police Station and realise that you have not searched his house for outstanding property. You spend the next hour and a half at the station completing forms and documenting the prisoner.
  It may have been more appropriate to search the house when you arrested Peter. | **125**

What powers do you have to conduct a house search?

....
....
....
....
....

▶128

¹¹⁷▶**126** | Where an arrestable offence has been committed, any person may arrest without warrant anyone whom he or she has reasonable grounds for suspecting to be guilty of it. | **126**

Are his answers alone sufficient to make an arrest?

....
....
....

In the circumstances that we have here you have acted reasonably. However, when the youth is arrested, he explains that the reason that he was not being completely honest with you is that he has just left his girlfriend's house as her parents arrived. He is not liked by them and did not want you checking where he had been because it would have got her into trouble.

What other evidence might have been available to you before you made the arrest and how could you have obtained it?

....
....
....
....
....

▶129

⁵⁵▶**127** | On this occasion, the youth you have stopped is innocent and you have made an unnecessary arrest. He is now going to make a complaint against you for unlawful arrest. | **127**

▶122

make another choice

$^{119}_{125}$►**128**   Section 32 PACE provides you with a really useful power. It can save you hours of time. Section 32(9)(b) also allows you to seize anything, other than an item subject to legal privilege, if you have reasonable grounds for believing that it is evidence of *an offence* or has been obtained in consequence of the commission of an offence.

This goes beyond the offence for which you are investigating.

What are the conditions under which you can conduct this search and what are the limitations of any search undertaken?

....

....

....

....

►111

$^{120}_{122}_{126}$►**129**

What are the circumstances under which you can conduct a PACE stop and search?

....

....

....

....

►35

^{88}►**130**   Code C, para. 11.1A and para. 11.1 outline what an interview is and where it should take place. Under the circumstances here, if you interview Peter at his home after arrest, you will render the evidence obtained as a result of the interview inadmissible.

This goes to prove that compliance with PACE and the Codes of Practice is essential in all investigations. What may appear to be a minor issue could end up with you losing an otherwise watertight case.

►33

make another choice

^{93}►**131**   You now have to interview Peter Grey.

►17

^{50}►**132**   You commence house to house and check the neighbours either side. The PC completes the crime report and brings it to you. She then starts to carry out a more thorough search of the house and grounds.

►78

^{77}►**133**   You don't expect me to believe that this is what you would really do?

►77

make another choice